PLACING NATURE ON THE BORDERS OF RELIGION, PHILOSOPHY AND ETHICS

The natural world has been "humanized": even areas thought to be wilderness bear the marks of human impact. But this human impact is not simply physical. At the emergence of the environmental movement, the focus was on human effects on "nature." More recently, however, the complexity of the term "nature" has led to fruitful debates and the recognition of how human individuals and cultures interpret their environments.

This book furthers the dialogue on religion, ethics, and the environment by exploring three interrelated concepts: to recreate, to replace, and to restore. Through interdisciplinary dialogue the authors illuminate certain unique dimensions at the crossroads between finding value, creating value, and reflecting on one's place in the world. Each of these terms has diverse religious, ethical, and scientific connotations. Each converges on the ways in which humans both think about and act upon their surroundings. And each radically questions the damaging conceptual divisions between nature and culture, human and environment, and scientific explanation and religious/ethical understanding. This book self-consciously reflects on the intersections of environmental philosophy, environmental theology, and religion and ecology, stressing the importance of how place interprets us and how we interpret place. In addition to its contribution to environmental philosophy, this work is a unique volume in its serious engagement with theology and religious studies on the issues of ecological restoration and the meaning of place.

Transcending Boundaries in Philosophy and Theology

Series editors:
Martin Warner, University of Warwick
Kevin Vanhoozer, Wheaton College and Graduate School

Transcending Boundaries in Philosophy and Theology is an interdisciplinary series exploring new opportunities in the dialogue between philosophy and theology that go beyond more traditional 'faith and reason' debates and take account of the contemporary reshaping of intellectual boundaries. For much of the modern era, the relation of philosophy and theology has been conceived in terms of antagonism or subordination, but recent intellectual developments hold out considerable potential for a renewed dialogue in which philosophy and theology have common cause for revisioning their respective identities, reconceiving their relationship, and combining their resources. This series explores constructively for the 21st century the resources available for engaging with those forms of enquiry, experience and sensibility that theology has historically sought to address. Drawing together new writing and research from leading international scholars in the field, this high profile research series offers an important contribution to contemporary research across the interdisciplinary perspectives relating theology and philosophy.

Also in this series

Beyond Fideism
Negotiable Religious Identities
Olli-Pekka Vainio

Talking about God
The Concept of Analogy and the Problem of Religious Language
Roger M. White

Kierkegaard and Levinas
The Subjunctive Mood
Patrick Sheil

Kant and Theology at the Boundaries of Reason
Chris L. Firestone

Placing Nature on the Borders of Religion, Philosophy and Ethics

Edited by

FORREST CLINGERMAN and MARK H. DIXON
Ohio Northern University, USA

Routledge
Taylor & Francis Group

LONDON AND NEW YORK

First published 2011 by Ashgate Publishing

Published 2016 by Routledge
2 Park Square, Milton Park, Abingdon, Oxon OX14 4RN
605 Third Avenue, New York, NY 10017

First issued in paperback 2021

Routledge is an imprint of the Taylor & Francis Group, an informa business

British Library Cataloguing in Publication Data
Placing nature on the borders of religion, philosophy and
   ethics. -- (Transcending boundaries in philosophy and theology)
   1. Philosophy of nature. 2. Human ecology--Religious
   aspects. 3. Environmental ethics. 4. Environmental psychology.
   I. Series II. Clingerman, Forrest. III. Dixon, Mark H.
   201.7'7-dc22

Library of Congress Cataloging-in-Publication Data
Placing nature on the borders of religion, philosophy, and ethics / [edited by] Forrest
Clingerman and Mark H Dixon.
      p. cm. -- (Transcending boundaries in philosophy and theology)
   Includes index.
   ISBN 978-1-4094-2044-6 (hardcover)
1. Ecology--Religious aspects. 2. Ecology--Philosophy. I. Clingerman, Forrest. II.
Dixon, Mark H.

   BL65.E36P63 2011
   205'.691--dc22
                                                           2011005819

ISBN 13: 978-1-03-224328-3 (pbk)
ISBN 13: 978-1-4094-2044-6 (hbk)

DOI: 10.4324/9781315600642

Contents

List of Contributors

Forrest Clingerman, Associate Professor of Philosophy and Religion, Department of Philosophy & Religion, Ohio Northern University.

Mark H. Dixon, Associate Professor of Philosophy, Department of Philosophy & Religion, Ohio Northern University.

Martin Drenthen, Assistant Professor, Faculty of Science, Radboud University of Nijmegen, Nijmegen, Netherlands.

James Janowski, Associate Professor of Philosophy, Department of Philosophy, Hampen-Sydney College.

William R. Jordan III, Director, The New Academy for Nature and Culture and Co-Director, DePaul University Institute for Nature and Culture.

Todd LeVasseur, Ph.D. candidate, Department of Religion, University of Florida. Adjunct Instructor of Religious Studies, College of Charleston.

David C. McDuffie, Ph.D. candidate, Department of Religious Studies, University of North Carolina, Chapel Hill.

Sarah Morice-Brubaker, Assistant Professor in Theology, Phillips Theological Seminary.

Sampson M. Nwaomah, Associate Professor of New Testament and Mission Studies, Department of Religious Studies, Babcock University, Ilisan-Remo, Ogun State, Nigeria.

Jonathan Parker, Ph.D. candidate, Department of Philosophy & Religion Studies, University of North Texas.

Anna L. Peterson, Professor of Religion, Department of Religion, University of Florida.

Páll Skúlason, Professor of Philosophy, Faculty of History and Philosophy, University of Iceland, Reykjavík, Iceland.

Daniel T. Spencer, Associate Professor of Environmental Studies, Department of Environmental Studies, the University of Montana.

H. Peter Steeves, Professor of Philosophy, Department of Philosophy, DePaul University.

David Utsler, Ph.D. candidate, Department of Philosophy & Religion Studies, University of North Texas.

Mélanie Walton, Adjunct Professor of Philosophy, Department of Philosophy, Kent State University.

A. James Wohlpart, Associate Dean, College of Arts and Sciences and Professor of English, Department of Language and Literature, Florida Gulf Coast University.

Preface

Convergence

As new problems, issues, and knowledge arise in the world, how does one build or create an intellectual discipline to confront these challenges? This might appear to be an academic question, though nevertheless it is pressing. The more common practice since the Middle Ages has been to isolate a more-or-less cohesive area within an existing discipline and create a sub-discipline in which researchers then can specialize. Specialization is seen as a desirable aim and, as a consequence, disciplines have become ever more narrow in their focus and content. In this scenario, in order to resolve a question scholars attempt to determine which domain best addresses the issue and the answer is presumed to deepen the scholarship within that discipline. While some disciplines might share some similarities—theology and philosophy are an obvious example of such a shared interest—it seemed clear when one's investigation was theological, and when it was philosophical.

As the world changes from the certainty of the twentieth century to the challenging uncertainty of the twenty-first, such insular boundaries seem less obvious. Perhaps we might even be apprehensive that knowledge—at least when understood as separate disciplines with little intercommunication—is about to collapse. In the complex interactions between social, intellectual, and individual life, scholars have begun to see the limitations that traditional boundaries between disciplines impose on innovative research. What this means is that, since there are inevitable similarities in content between separate disciplines, one can treat the area where the disciplines intersect or converge as an intellectual discipline in its own right. Though the historical response has been to treat these borders with suspicion, there can be no doubt that these convergence points are real—not only as intellectual constructs, but as existentially important facets of human life.

This volume is about one such convergence point: place. A rather common and unpretentious word to possess such a crucial theoretical position, nevertheless the concept "place" has a fundamental role in the arts, humanities, and sciences. As beings with physical bodies, our embodiment implies our emplacement. At each moment in our lives we are somewhere, at some place, at some time in the physical universe. This is a realization that most Western philosophical systems have seen as being too inconsequential to have much philosophical interest. There is a separate tradition though—one that traces its origins through Hegel and Husserl to Heidegger—the phenomenological tradition—that sees this realization as *the* essential philosophical insight into human nature.

This realization is more than an abstract philosophical recognition—it has more immediate implications as well. As Heidegger's famous maxim has it, "to be is to dwell" and to dwell means to change the environment and landscape in which we dwell. Thus through our decisions and actions we alter the places where we live—sometimes these changes are beneficial, *at least to* human beings, and at other times less beneficial, *even to* human beings. In turn, the places we inhabit influence us—who we are, what we do and how we conceive of ourselves in relation to other human beings, non-human animals and the wider natural environment. These are complex influences and interactions. Indeed the complexities are such that it is often difficult (perhaps even impossible) to demarcate the point between human influence on the place and place's influence on human beings. That the influence is mutual is uncontroversial; the precise mechanisms and degrees, however, require cautious thought and consideration.

Thus what it means to dwell—for Heidegger as well as for theologians, philosophers, and other scholars—is not simply a question about how we are located in place. The very concept "place" determines how we interpret who and what we are—how we emplace ourselves within the world. Furthermore it is impossible to describe or to analyze our abilities to understand place and ourselves through one intellectual discipline. *A more robust thinking is needed—a thinking that is theological, philosophical, scientific and, in the end, "human" in the truest sense.* When we consider such catastrophes as the 2010 oil spill in the Gulf of Mexico, the enormous Pacific Trash Vortex or even more gradual destruction such as global climate change and loss of biodiversity, there can be no doubt that there are clear problems with how human beings, both as individuals and societies, understand and negotiate "place" and its role in their lives. Thus investigations and explorations into a concept—such as "place"—that represents a convergence point between disciplines represents a clarion call to disciplines across the intellectual and academic matrix to themselves converge upon this area. Such a convergence is essential: a fact that the chapters of this volume both presuppose and further develop.

Most academic disciplines bring some particular insight into the importance that place has—both in the abstract and in the concrete. So what happens when we situate or place these disciplines into conversation with each other? What can the scientist tell the philosopher about place? What can the philosopher or theologian tell the artist? What can the artist tell the scientist? What seems undeniable is that such conversations can enrich our knowledge about ourselves and our interactions with the natural environment. But if the present book rests at the borders of religion, ethics, and philosophy—as its title suggests—it does so in order to seek a new way forward. Given the pressing nature of environmental problems, the authors of this volume are suggesting ways of revitalizing how we know the world. Thus while the present collection emphasizes theology, philosophy, and ethics, it does so only as a starting point with the desire for a more wide-ranging dialogue and with the acknowledgement of the limitations of these fields. And while these chapters are academic in nature, it is also hoped by editors and authors alike that the collegial discussion provided in this book can be considered a step toward public scholarship on place, environment, and a more humane future.

"Placing" Nature

With one exception all the chapters in this volume began life as presentations at the conference Recreate, Replace, Restore: Exploring the Intersections Between Meanings and Environments, which was held at Ohio Northern University in April 2009. The aim of the conference was to bring together scholars across academic disciplines to discuss the interactions between their disciplines as concerns "place." Thus these chapters use various perspectives, disciplines, and methodologies to explore the concept "place" and all the complexities that it implies. Despite their differences, though, there are certain themes that bind these chapters together, in particular all the chapters explore the area where religion, ethics, natural sciences, and place meet. The initial discussions in Ada, Ohio (just south of the Great Black Swamp, as participants were told in the initial orientation to the locale given by Robert Verb of ONU's Department of Biological and Allied Health Sciences) demonstrated how scholars that represented disciplines across the humanities and the sciences, that came from four continents and eight countries, and that brought different intellectual questions and problems, could focus their research on a common question—"place"—and together create a genuine intellectual dialogue that was able to move between and across conventional boundaries.

The volume begins with the one non-conference chapter, the Icelandic philosopher Páll Skúlason's "On the Spiritual Understanding of Nature." In this chapter Skúlason explores the deep and spiritual connection that place can have on us through his own experiences at Askja, a volcano in Iceland's interior. What this connection reveals to us, Skúlason suggests, is that we are "earthlings", i.e. bound to the earth at so deep a level that it becomes a fundamental premise in our lives. To unravel this premise Skúlason appeals to the "numinous," a concept that Rudolf Otto explicates in considerable detail in *The Idea of the Holy*.

Skúlason's chapter represents, as it were, a conceptual nexus that introduces the various ideas and connections that the subsequent chapters explore in some detail. It is perhaps serendipitous that these chapters *organize themselves* into separate categories—chapters that explore particular places and the influences that human beings have had on them, and chapters that represent more theoretical reflections on the influence that place has on human life and purpose.

The section that includes those chapters that focus on particular places—*Restoring Place and Meaning*—illustrates the complex interactions between human beings and the natural environment and what happens when human actions focus on human concerns but ignore environmental concerns. That such actions do cause environmental destruction is uncontroversial. What these chapters disclose though is that these actions can also devastate how we see ourselves and define our, individual and social, role and purpose within the wider natural environment.

Environmental destruction then undermines the distinction between others and us in such a fundamental manner that in order to save ourselves it becomes essential to save the environment as well. Thus the chapters in this section grapple with the problems that environmental restoration poses to human beings as well as to natural and built environments. In other words, the authors seek to get the reader's hands dirty, as it were, by illuminating some of the theoretical and practical issues involved in restoring actual places from the ground of individual and communal human lives.

The chapters in the final section—*Recreating Place, Reconnecting with Others*—explore in a more hermeneutical and critical manner the complex interrelations between human beings and place. This complements the previous section by deepening the discussion of what meanings we bring to and receive from the places in which we dwell. As the section title suggests, when human beings acknowledge their role in environmental destruction and engage in reconstructive activities the end result is that we in some sense recreate ourselves as well as the environment. The ties that bind human beings to the environment mean that unilateral influence is impossible. To be embodied is to be emplaced, to be human is to dwell—thus actions that change the environment change human beings as well. These changes are more than scientific, there are theological, ethical, aesthetic and other changes as well—changes that operate on, and at, deep personal and social levels. The chapters in this section explore then the human–environment connection through changes in the theological, ethical, aesthetic and other "lenses" brought to bear on human and natural places.

At the Recreate, Replace, Restore conference it became obvious that there was both a need and a desire to begin to dismantle the traditional intellectual boundaries and to promote and advance less discipline-bound dialogues and methodologies. It is our conviction that these chapters, in their attempts to "place nature" outside traditional intellectual boundaries, will serve both that need and that desire.

Acknowledgments

This volume represents a collaborative endeavor, between the editors to be sure, but also between all the participants at the Recreate, Replace, Restore conference whose chapters appear here. We wish to thank them all—their participation and presentations made this an excellent conference.

We also want to acknowledge the Working Group on Religion, Ethics and Nature (WGREN) Conference Committee (Fern Albertson, Wayne Albertson, Bill Fuller, Jay Mager, and Suzanne Morrison) who organized the conference. Jay Mager was a valuable conversation partner in our planning, particularly in framing the conference themes from a scientific perspective. Kelsey Kern and Phoebe Stroede, the two student members on the Conference Committee, deserve particular mention—their initiative and assistance both before and during the conference was invaluable. We also wish to acknowledge Naoko Sasaki's assistance with the conference paper selection process and their organization into the conference sessions—her insights made this process much less arduous. As with all academic conferences logistics is a critical concern, but one we were able to leave in Axel Brandt's, Sara Zeiger's, and Jenny Gargac's capable hands. Finally, we were fortunate to have the support of the Metanexus Institute, whose Local Societies Initiative grant made WGREN possible.

At Ohio Northern University, the Office of the President, the Office of the Vice-President for Academic Affairs, the Office of the Dean of the Getty College of Arts and Sciences and the Department of Philosophy & Religion provided the support and encouragement that were essential to the conference's success. In particular we wish to express our gratitude to President Kendall Baker, Vice-President Anne Lippert, Dean Lisa Robeson and Dr. Raymond Person, Jr, the Chair of the Department of Philosophy & Religion.

Our thanks also to Kathleen M. Dixon whose assistance with the final read through was invaluable. Finally, we owe a debt of gratitude to Kevin Vanhoozer and Martin Warner, editors of the series Transcending Boundaries in Philosophy and Theology. Their interest in our project, as well as their insights and guidance at several stages, have made this a stronger volume.

Forrest Clingerman
Mark H. Dixon
Ada, Ohio, December 2010

To our respective families, Gail, Sabina and Asa Clingerman, and Kathleen Dixon, without whose support and encouragement this would have been impossible.

Chapter 1

On the Spiritual Understanding of Nature

Páll Skúlason

I

Some years ago I wrote a book which I called *Meditation at the Edge of Askja* where I tried to explain some of the thoughts that Askja, a volcano in Iceland, had awakened in me after my first visit there in 1994. When I was walking back from Askja, some very speculative questions were growing in my mind: how do totalities, self-contained wholes, come to be, and what kind of totalities are there? And how do connections come to be, and what kinds of connections are there?[1] Some months later, after having spent some months in Paris carrying with me a stone from Askja, I started reflecting upon my encounter with Askja and comparing it to my experience of the city of Paris.

Coming to Askja was for me like coming to Earth for the first time and discovering myself as an earthling: a being whose very existence depends on the Earth, a being who can only be itself by relating to this strange, overwhelming and fascinating totality, which is already there and forms an independent, objective, natural world. Askja symbolized for me an "objective reality, independent of all thought, belief and expression, independent of human existence."[2] Askja, I said, "is the earth itself as it was, is, and will be, for as long as this planet continues to orbit in space, whatever we do and whether or not we are here on this earth. Askja was formed, the earth was formed, long before we were created. And Askja will be here long after we are gone."[3] For me it was suddenly evident that everything else, every totality or connection that I could discover or imagine, could only be a reality because of its relation to the natural reality which Askja symbolized so marvelously. By contrast Paris, that fascinating city, was a man-made totality of an entirely different kind, but dependent upon the existence of nature as an autonomous and independent reality.

---

[1]   These questions can clearly be attached to everything we encountered, not only to a visible phenomenon like a volcano or a man-made phenomenon like a city or a house, but also to invisible beings like a state or a nation or the self that each of us carries with him or her.

[2]   Páll Skúlason, *Meditation at the Edge of Askja* (Reykjavík: University of Iceland Press, 2004), p. 21.

[3]   *Ibid.*

My thesis in the former book—if one can call it a thesis—was, in short, that to be "an earthling is to feel one's life to be bound to the earth, or deriving from it, to feel the earth to be the fundamental premise of one's life." I argued for this thesis in the following way: "I am I, you are you, and we are we because we place ourselves, are and cannot be what we are except in the face of Askja (or other, comparable, symbol of the earth), to which we can turn again and again, if not in actuality, then in our thoughts. We stand upon the earth—build, work, and destroy it, if it comes to that—because we are born to the earth and can only find ourselves in relation to it, in the light of it or in its embrace. The earth is thus the beginning and the end of all our feeling for reality as a unified totality, and thus of all our feeling for ourselves as inhabitants of the world. The earth is the premise of our being ourselves, of our existing together and being aware of ourselves."[4]

In the book I am referring to, my main aim was to illustrate this spiritual understanding of the Earth in all its natural glory as an independent reality into which we are born and with which we engage in a complex relationship.[5] I call this understanding "spiritual" because it has to do with our feelings and our sense of value, of belonging to and being separated from, of being afraid and being fascinated. It is an understanding that has to do with religion rather than science, magic rather than technology. Its religious connotation does not make it any the less important. On the contrary, it reveals a problem which is not primarily a scholarly problem, but concerns the most basic connections of the mind to reality, and all the uncertainty and insecurity that pertain to our relations as sentient and reflecting beings to natural reality itself.

In the present chapter I want to explore some aspects of this spiritual understanding of nature, which I tried to illustrate in the former book. Such an understanding does not of course replace the scientific understanding of nature to which we are all accustomed and which is the basis for our technological powers. But it may help us to see and to think about issues that we might otherwise overlook, issues that concern how we are to realize the basic values and ultimate goals of our existence. It is the function of civilization, or if you like, of culture, of politics and of economics, to secure these values and goals. But a civilization may also conceal the issues that need to be dealt with, and the question that confronts humanity now is whether we recognize and work toward what ultimately matters and whether we may have to change radically the ways in which we are developing our civilization.

One way of criticizing our present civilization is to say that it has made the value of *efficiency*, which pertains to certain means that are at our disposal, into an ultimate value, that is to say into a sacred value, and that this *spiritual* error, if I may call it that, is leading us astray. This criticism, which we can find in the writings of many thinkers of the last century, among them Heidegger and Wittgenstein, must certainly be taken seriously. But if this error is to be corrected we need to

---

4   *Ibid.*, pp. 21–3.
5   See *ibid.*, p. 27.

develop a proper understanding of the reality of which we are a part and which may enable us to develop more appropriate relations to our fellow beings and to the world as a whole. I believe that both Heidegger and Wittgenstein, along with many other scholars, poets and philosophers, contributed significantly to such an understanding—a spiritual understanding. One of their lessons is that we should not expect to find solutions to our existential worries in a grandiose theory but rather in a humble way of reflecting upon our own experience and what others can tell us about their experience of the world.

II

Bearing this in mind, let us return to the experience of Askja with the help of another story of a first encounter with that magnificent volcano. In the year 1923 an Icelandic scientist, Pálmi Hannesson, came to Askja for the first time. I discovered his account of his first encounter with Askja by chance, only a few months ago. Hannesson speaks of walking toward Askja, enjoying the wonderful views of the Icelandic highlands, when we enter his adventure:

> After a short while I arrived at the edge of a sheer cliff and saw Askja lying before me—or was it I who lay before Askja? At my first glimpse, I had to look away. Nothing like that has happened to me before or since, to be struck dumb by landscape. But there is some magic attached to Askja, some awesome, disturbing force that took me unawares and that I could not at first withstand, there in my solitude. I have never seen anything as astonishing or powerful. It was as if the magnificent view which I had enjoyed just a moment before had been erased from my mind, and with the terror of animate flesh, I was confronted by this awesome wonder of inanimate nature. There is no hope of describing Askja in any meaningful way. Who can describe a great work of art? Words and images are like the mere clanging of metal or the beating of a bell. And the same applies to any attempt to describe Askja.[6]

Two essential elements are clear in this account. The first is the overwhelming impression made by Askja upon the mind of the perceiver: he has to look away, he cannot face this reality. He finds himself "struck dumb" (in fact the metaphor is melting), he was unprepared and had no defense against the awesome, disturbing force of Askja. Everything else fades away and he is experiencing the "terror of animate flesh" ("the fear of the living flesh") in face of this mysterious and tremendous power of "inanimate nature." The second aspect is his "speechlessness": he is deprived of the power to describe or explain what he is discovering, but nevertheless tries to say something.

---

[6]   Pálmi Hannesson, *Landið okkar* (Reykjavik: Bókaútgáfa Menningarsljóðs, 1957), p. 109.

What Pálmi Hannesson is telling us about is an experience of what Rudolf Otto has described in his book, *The Idea of the Holy*, as the experience of the *numinous*.[7] "Numinous" is a word coined by Otto in order to denote what we take to be "holy" or "sacred," but without any ethical or rational connotation. The numinous is a specific category of value which can only be encountered in a special experience that gives rise to a unique state of mind: "This mental state is perfectly *sui generis* and irreducible to any other; and therefore, like every absolutely primary and elementary datum, while it admits of being discussed, it cannot strictly be defined."[8] Otto's book, which was first published in English in 1923 (the year that Hannesson visited Askja!), is an attempt to discuss and clarify the nature of the numinous "by means of the special way in which it is reflected in the mind in terms of feeling."[9] In this experience we are, according to Otto, "dealing with something for which there is only one appropriate expression 'mysterium tremendum'."[10] He distinguishes between three basic elements in such an experience. First is the element of Awefulness, which is the tremor, the fear, the dread, or the "symptom of 'creeping flesh'."[11] The "Wrath of Yahweh" in the Old Testament has something to do with this. "There is something very baffling in the way in which [the 'Wrath of Yahweh'] 'is kindled' and manifested. It is, as has been well said, 'like a hidden force of nature', like stored-up electricity, discharging itself upon anyone who comes too near."[12] The second element, closely connected to the first, is that of "Overpoweringness" or what Otto names *majestas*. It is the

---

[7]   A colleague and friend, Mikael M. Karlsson, who translated *Meditations at the Edge of Askja* into English, pointed out to me that the experience I described in that text had certain similarities to what Rudolf Otto was discussing in his book as the experience of the numinous. The similarities are even more striking when it comes to Hannesson's account of his experience of coming to Askja.

[8]   *The Idea of the Holy* (Oxford: Oxford University Press, 1958 [1917]), p. 7.

[9]   *Ibid.*, p. 12. The feeling which Otto describes and discusses at length is that of dependence: "It is the emotion of a creature, submerged and overwhelmed by its own nothingness in contrast to that which is supreme above all creatures" (p. 10).

[10]   *Ibid.*, p. 10. Otto describes it in the following way: "The feeling of it may at times come sweeping like a gentle tide, pervading the mind with a tranquil mood of deepest worship. It may pass over into a more set and lasting attitude of the soul, continuing, as it were, thrillingly vibrant and resonant, until at last it dies away and the soul resumes its 'profane', non-religious mood of everyday experience. It may burst in sudden eruption up from the depths of the soul with spasms and convulsion, or lead to the strangest excitements, to intoxicated frenzy, to transport, and to ecstasy. It has its wild and demonic forms and can sink to an almost grisly horror and shuddering. It has its crude, barbaric antecedents and early manifestations, and again it may be developed into something beautiful and pure and glorious. It may become the hushed, trembling, and speechless humility of the creature in the presence of—whom or what? In the presence of that which is a mystery inexpressible and above all creatures" (pp. 12–13).

[11]   *Ibid.*, p. 16.

[12]   *Ibid.*, p. 18.

feeling of an absolute dependency: "Thus, in contrast to 'the overpowering' of which we are conscious as an object over against the self, there is the feeling of one's own submergence, of being but 'dust and ashes' and nothingness."[13] The third element is that of the "energy" or urgency of the numinous object; this element is already involved in those of *tremendum* and *majestas*. Everywhere it "clothes itself in symbolical expressions—vitality, passion, emotional temper, will, force, movement, excitement, activity, impetus."[14]

For Otto the category of the Holy or the numinous object is a purely a priori category in the Kantian sense. It is neither a supra-natural object nor can it be reduced to sense-experience or said to be evolved from some sort of sense-perception.[15] "It issues from the deepest foundations of cognitive apprehension that the soul possesses, and, though it of course comes into being in and amid the sensory data and empirical material of the natural world and cannot anticipate or dispense with those, yet it does not arise *out* of them, but only *by their means*."[16]

To come back to Hannesson's experience of Askja, and to mine as well, it is obvious that both these personal experiences bear all the marks of "numinous consciousness" as if Askja had been an occasion for us to discover the numinous object. Askja was, to use Otto's terms, "the sensory data and empirical material of the natural world" by means of which we were filled with the emotion of the numinous. Askja had awakened in us "a numinous consciousness" which, according to Otto, points to "a hidden substantive source, from which the religious ideas and feelings are formed, which lies in the mind independently of sense-experience."[17]

Now my question is: what lessons concerning our understanding of nature and of our relationship with Earth as our home in the natural world can be drawn from this experience of the numinous awakened by the volcano Askja?

---

[13]   *Ibid.*, p. 20.

[14]   *Ibid.*, p. 23. According to Otto this aspect of the numinous, the element of energy, "reappears in Fichte's speculations on the Absolute as the gigantic, never-resting, active world-stress, and in Schopenhauer's daemonic 'Will'" (p. 24). Then Otto makes a remark which gives us an important indication about his way of thinking: "At the same time both these writers are guilty of the same error that is already found in the myth; they transfer 'natural' attributes, which ought only to be used as 'ideograms' for what is itself properly beyond utterance, to the non-rational as real qualifications of it, and they mistake symbolic expressions of feelings for adequate concepts upon which a 'scientific' structure of knowledge may be based" (p. 24) The numinous or the Holy object which the mind discovers as "the mysterium tremendum" is not to be explained by a metaphysical theory about a supreme power that some philosophers may imagine as that to which everything in the universe must ultimately relate.

[15]   See *ibid.*, p. 112.

[16]   *Ibid.*, p. 113.

[17]   *Ibid.*, p. 114.

I will proceed from the abstract to the concrete and start with a reflection about experience in general and the way in which it connects with reality, then proceed to a reflection on the experience of the numinous as such and what it means for our understanding of reality, and finally discuss the meaning of the specific experience of Askja for our actual understanding of Nature and Earth.

III

In order to make clear to ourselves what experience means for us I think it may be useful to take into account what Hegel has in mind when he discusses the concept of experience in the introduction to his *Phenomenology of Spirit*.[18] Hegel explains in this introduction how he understands the logic implicit in the process by which we gain experience of the world and of ourselves. The experience he has in mind is not only sense-perception, although he sees that as the very beginning of the process of human experience. The experience Hegel is dealing with is that of the conscious being that is at the same time conscious of the reality outside itself and of its own internal reality. In short, the process of experience that Hegel describes implies two steps. The first one is the discovery of something that exists outside of, and independently of, our consciousness; the second step is the arrival on the scene of a new object, created by the encounter of human consciousness with the first object. The first object, which existed only in itself and independently of our consciousness, still exists, of course, in itself but now also for our consciousness.

Hegel points out that there is a certain ambiguity here concerning the truth. Our consciousness has two objects: the reality which is there and the reality which is there for consciousness. Now the second object is apparently only the knowledge we have of the first object. According to our ordinary understanding of experience, we gain this knowledge by correcting an imperfect notion we had beforehand of the first object. But what we do not realize, Hegel remarks, is that the encounter transforms both our consciousness and the first object and generates a second object for our consciousness, namely experience itself as a unity of what is out there and what is in our mind. So when we are dealing with our own personal experience we are always wondering which part is determined by external factors and which by our own mind. For Hegel, we should approach experience itself as a creative process, a process in which new objects for our consciousness are constantly being formed. That allows for new relations between the world and ourselves and thus for a new understanding of reality. It is thus that we can understand the meaning of what we experience. The *Phenomenology of Spirit* is the story of how new objects, ever richer in meanings, come on the scene, starting with the first objects of sensation here and now and ending with the concept of Absolute knowledge as

---

[18]   First published in German in 1807. I am inspired here by Heidegger's essay, "Hegels Begriff der Erfahrung" ("Hegel and his Concept of Experience"), in *Holzwege* (Frankfurt am Main: Klostermann, 1950), pp. 105–92.

an object for our consciousness to reflect upon (and passing through interesting stages of experience where objects conveyed by concepts of self and reason play a major role in organizing our thinking about reality).

To take an example from our previous discussion of the numinous, it is clear that the numinous does not exist either in nature or in the mind but is brought into reality by their encounter with one another, under special circumstances. But in our ordinary way of thinking we have a tendency to place the numinous outside our consciousness, in a reality that is either natural or supra-natural. To do that we need a metaphysical theory about Nature and what may lie beyond it to explain the position of the numinous (Otto mentions Fichte's and Schopenhauer's theories). But there is also another way of locating the numinous, namely by placing it among the pure a priori categories of the transcendental mind which Kant considered to be the ultimate conditions of possibility for all experience. The first option is some kind of realism or naturalism which we all have a natural tendency to follow. The second option is a transcendental idealism which owes much to Plato, and this is the one that Otto favors.

The third option is the one that Hegel opened up for us in his *Phenomenology*, i.e. that of understanding the numinous, and all other symbols or expressions of our experience, as objects that owe their existence to our real encounters in the world as conscious beings. Following this option, we have to learn to approach and understand reality as the encounter of consciousness and the world (which initially means the natural world or simply nature). This encounter embraces all relations or connections where mind and nature meet in fact. But that does not mean that all encounters are equivalent. On the contrary, if *experience* teaches us anything it is that there is a difference between one encounter and the next. And if *reflection* or thinking teaches us anything it is that we can account for these differences in unforeseeable ways in our linguistic and symbolic expressions. I take our *reflective* capabilities to be closely related to our *creative* capabilities and to be unlimited in their capacity to produce things or ideas that make sense or that do not—or seem to us not to—make any sense. And what makes sense at one specific moment may not make any sense in other circumstances. But even though we, as conscious beings, are capable of producing an infinity of objects for our consciousness to rejoice in or get excited about, our real capacities to understand our experiences of the world are constrained by the cultural systems which have been created in order to ensure our spiritual and physical security.

Let me explain. We need three things to survive in the world. First, physical sustenance (shelter, food and security); second, institutions to organize our social relations; and, third, ideas to keep track of our thinking. Former generations have already established a variety of systems to guarantee the continuation of our species on Earth: systems for the production and distribution of worldly goods, systems of governance and a division of powers, and systems of ideas which explain to us how and why things are as they are in the world.

Now these systems may serve their purpose well or badly, but the basic problem with them is that they come between our minds and the reality out there and may completely blind us to the extraordinary dimensions of the world that lies beyond all the systems we can possibly invent or imagine. It is as if these systems were put in place to protect us from reality itself by controlling and deciding in advance the meaning of our future experiences. So in order to set our mind free we must break away from these cultural systems in order to develop our own thinking and to experience the world and ourselves on the basis of our personal relations with reality.

IV

Here we come to the experience of the numinous, which is an experience that does not fit into any system of ideas but shows the superficiality and smallness of all such systems. Let us now reflect upon this experience as such, that is, without taking into account which specific external object gives rise to it or by what means it was brought about (for instance by a drug).

From Otto's analysis it is clear that the numinous consciousness is not related to any specific external circumstances, and he sees the numinous rather as an extraordinary phenomenon or object which we may discover and experience in a great variety of situations. In fact there seems to be something about the world which makes it an occasion for us, at least from time to time, to have the emotional experience of the numinous in the sense that Otto gave it, i.e. something sacred or holy but without any moral or rational content. For Otto this special emotion can only be brought about because in the depths of our soul we are endowed with the extraordinary capability to discover the numinous.[19] According to Otto we should think theoretically of the origin of the numinous as "'pure reason' in the profoundest sense, which, because of the 'surpassingness' of its content, must be distinguished from both the pure theoretical and the pure practical reason of Kant, as something yet higher or deeper than they."[20]

I am not going to dispute this Kantian way of thinking, but it seems to me much more interesting to concentrate on the fact that there is something disturbing in our relationship with reality; something which occasionally makes us experience the world as totally beyond all comprehension. The emotion of the numinous or, if you prefer, the numinous consciousness, is a clear revelation of this basic fact

---

[19]   A bishop in Iceland was arguing at Easter that the idea of God is hidden in [lies within] every human soul, but that we have ourselves to make the effort of activating it in daily life and experience. Otto would not have called the numinous God although he was a Christian. It is rather a pure a priori category of reason in the Kantian sense. "The ideas of the numinous and the feelings that correspond to them are, quite as much as the rational ideas and feelings, absolutely 'pure', and the criteria which Kant suggests for the 'pure' concept and the 'pure' feeling of respect are most precisely applicable to them" (Otto, *The Idea of the Holy*, p. 112).

[20]   *Ibid.*, p. 114.

about our connection to reality, to the world and to ourselves. Of course, we do not usually dwell on this disturbing experience which people may experience and express quite differently. I have always found Albert Einstein's account of this experience very much to the point: "The most beautiful thing we can experience is the mysterious. It is the source of all true art and all science. He to whom this emotion is a stranger, who can no longer pause to wonder and stand rapt in awe, is as good as dead: his eyes are closed."[21] I take it for granted that the different religions of the world offer various ways of articulating this spiritual understanding in order to make it possible for us to secure our mental ties to nature and between ourselves. The word "religion" comes from *re-lier*, to bind together, and this simple meaning is of utmost importance, it tells us what religion is all about. But if there is a need to bind together, or to relate to something, it is because there is *separation*: I discover myself separated from everything else, even from myself as a natural being.

V

This separation is reflected in the first idea that springs to mind when discovering Askja. It is what Hegel would call the absolute *exteriority* of nature to the human mind. This is what Otto describes as "the wholly other": "The truly 'mysterious' object is beyond our apprehension and comprehension, not only because our knowledge has certain irremovable limits, but because in it we come upon something inherently 'wholly other', whose kind and character are incommensurable with our own, and before which we therefore recoil in a wonder that strikes us chill and numb."[22] We are—remember—coming to Earth for the first time.

But this discovery of nature as a pure *exteriority* is at the same time a discovery of ourselves as a pure *interiority*, as beings aware of the world and aware of themselves as being aware, a discovery which Pascal describes in his *Pensées* in the following way: "Man is but a reed, the most feeble thing in nature; but he is a thinking reed. The entire universe need not arm itself in order to crush him; a vapor, a drop of water, suffices to kill him. But even if the universe were to crush him, man would still be more noble than that which kills him, because he knows that he dies, and knows the advantage that the universe has over him; the universe knows nothing of this."[23]

---

[21]   Michael White and John Gribbin, *Einstein: A Life in Science* (New York: Dutton/ Penguin Books, 1994), p. 262.

[22]   Otto, *The Idea of the Holy*, p. 28.

[23]   "L'homme n'est qu'un roseau, le plus faible de la nature, mais c'est un roseau pensant. Il ne faut pas que l'univers entier s'arme pour l'écraser; une vapeur, une goutte d'eau suffit pour le tuer. Mais quand l'univers l'écraserait, l'homme serait encore plus noble que ce qui le tue, parce qu'il sait qu'il meurt et l'avantage que l'univers a sur lui;

Such is the first step of the experience in question. Nature—in this case the volcano Askja—imposes itself on us in such a way that we lose touch with all ordinary reality; we are struck dumb and become pure awareness.

Let us now consider the second step which is the formation of the content of this experience, its proper meaning. This is, in Hegelian terms, the second object which is generated by the first encounter. This second object is the discovery that my very existence consists in connecting with nature, in establishing an adventurous and creative relationship with her, to recognize myself as an earthling. Now we come to the thesis which I mentioned at the beginning of my chapter regarding what it is to be an earthling. It is "to feel one's life to be bound to the earth, or deriving from it, to feel the earth to be the fundamental premise of one's life." What matters is that we exist, each and every one of us, in our effort to relate to Nature, or more exactly to the powers which make Nature a reality for us. If this is true, then everything depends on how we—as perceiving, thinking and acting beings—establish and develop our personal relationships with the external world with its infinity of extraordinary things and beings and places.

VI

The question that still remains concerns the universal validity of this intimate relationship or, if you prefer, of the lesson that can be drawn from it. Is not the whole of my argument based upon romanticism, a desire to return to mother nature and forget all the ugliness and pollution of the so-called civilized world? And does not this romanticism originate in some kind of an irrational mysticism, which presupposes an unknown capability of the mind to relate to what is beyond human comprehension?

Now I could confess that I am not especially inclined to romanticism and that I do not believe in mysterious powers beyond human understanding although reason tells me that there must be powers in the universe that we do not know or understand. But I do not believe that the terms "romanticism" or "mysticism" are helpful to deal with what is at stake here. What is at stake is *understanding* the relationship we form in our experience and in our mind with nature and how we succeed in clarifying this experience to ourselves and others. Let us look at an example from the perspective of a scientist.

In his book *Gaia: A New Look at Life on Earth* and his latest book *The Revenge of Gaia*,[24] James Lovelock, a medical doctor and a scientist, explains splendidly how the metaphor of Gaia (originally a Greek goddess symbolizing Earth) helped him clarify his experience of our planet as a living being, that is as a self-regulating

---

l'univers n'en sait rien." Pascal, *Pensées et opuscules* (Paris: Hachette, 1963), section VI, paragraph 347.

[24]   James Lovelock, *Gaia: A New Look at Life on Earth* (Oxford: Oxford University Press, 1979); *The Revenge of Gaia* (New York: Basic Books, 2006).

whole that obeys its own values, rules and goals. For Lovelock it is clear that we need to develop a new spiritual understanding of Nature in order to prepare for what is to come; and to follow him we have to develop a way of thinking in which we accept a vocabulary which is appropriate for our relationship to nature in our real experiences of it beyond what we can grasp in our conscious thought. "Important concepts like God or Gaia are not comprehensible in the limited space of our conscious minds, but they do have meaning in that inner part of our minds that is the seat of intuition. Our deep unconscious thoughts are not rationally constructed; they emerge fully formed as our conscience and an instinctive ability to distinguish good from evil."[25]

I believe that Lovelock is right and that we have a long way to go to develop and understand our relationship with Nature and ourselves. Our basic ideas, which mediate our experiences and help make sense of our encounter with Nature and with ourselves, are not rationally constructed like our technological devices. In fact we are moved by ideas much more than we move them.

At the beginning of this chapter I reminded you of the idea of *efficiency* which our present civilization has elevated to a supreme value. One might say that the economic system of the world thrives on more and more efficiency in producing and distributing worldly goods among people. At the same time it is the same economic system which, according to many scientists, is by its efficient manner of using our earthly resources affecting the self-regulating system of Earth in ways that may lead to a disaster for all living beings, not only for us humans that may be responsible for this.

One of the most important issues of our times may be how to formulate and implement ideas which would make the economic system of the world function well without going against the interests of the self-regulating Earth as a whole.

VII

I would like to end this chapter by introducing an idea which might counter-balance the idea of *efficiency* as the regulative value of all our activities. The idea I have in mind has been with us from the beginning of this chapter and it is, I believe, the main lesson to be learned from the experience of nature I have been discussing. It is the idea of *wholeness*. It seems to me that all the other ideas which have been mentioned in connection with this experience relate to this basic idea of *wholeness*

---

[25]   Lovelock, *The Revenge of Gaia*, p. 177. Lovelock believes that we still have a long way to go to learn to establish a proper relationship to and understanding of the Earth as a living, self-regulating system: "Our religions have not yet given us the rules and the guidance for our relationship with Gaia. The humanist concept of sustainable development and the Christian concept of stewardship are flawed by unconscious hubris. We have neither the knowledge nor the capacity to be stewards or developers of the Earth than are goats to be gardeners" (p. 176).

or *totality*, if you prefer. The concepts of God and Gaia, to speak like Lovelock, are deeply connected to this idea. According to Christian belief, nature as a whole is God's creation and Gaia is obviously a metaphor which has the function of making us view and understand the Earth as a unique living being, integrating all other organisms and having the function of safeguarding the conditions of life within its boundaries.

The first instance of the experience of Askja as a microcosm of Nature in its totality is that of the *separation* between ourselves and nature, the second one is that of discovering our existence as consisting in our *relationship* with nature. The world of our experience is the totality of the connections that may develop between us and the powers of nature, and we need concepts that help us cope with the infinite variety of connections that there may be. If the concept of *efficiency* may be helpful in order to multiply these connections and make use of everything we encounter, the concept of *wholeness* may be helpful in order to see the unity and the diversity of all these connections and which of them are for the good and which of them are harmful not only for us, but also for all living beings and the Earth itself as our home in the universe. Of course, our capacity to understand totalities and connections is quite limited, but we can nevertheless put ourselves *to a certain extent* and *in imagination* in the place of all other creatures and even see things, as Spinoza said, *sub speciae aeternitatis*, from the perspective of eternity. God and Gaia are names for beings that we can imagine to have the most important perspectives which we need in order to view the reality of our experience as a whole. And we should never forget that our experiences are experiences of a reality which is forcing itself upon our mind and imposing the rules of the game much more effectively than we can ever do.

Our task must be to try to understand these rules and imagine all the possible moves that can be made in this game where life on Earth is at stake—a game which may be played by rules of which we do not have the slightest idea. But we have also invented our rules and one of them is that culture and nature should be kept apart, because they are different. On the one hand we have the visible reality which is offered to our senses, a volcano or a waterfall, on the other hand we have the invisible reality of meanings carried by our words, which name the realities we can perceive, their characteristics, our relations to them or whatever. *Nature* is the name for this reality out there independent of our consciousness, and *culture* is the name for the reality we make ourselves with rational concepts as the basic tools for both technological and ideological constructions of our civilization.

Apparently we live in two worlds: the world of Nature and the world of Culture. The idea of wholeness that I am forwarding defies this distinction. And all my arguments so far have been intended to show that this distinction does not make sense in the reality of our experience, which is an encounter between consciousness and nature and is the continued development of this encounter under new and unforeseeable circumstances. The question that arises continuously in the process of experience is: what is for the good and what is harmful for life; what is destructive, and what makes life flourish? As Wendell Berry says "the concept of health is rooted in the concept of wholeness. To be healthy is to be whole. The word *health* belongs to a family of words, a listing of which will suggest how far the consideration of health must carry us: *heal, whole, wholesome, hale, hallow, holy*. And so it is possible to give a definition to health that is positive and far more elaborate than that given to it by most medical doctors and the officers of public health."[26]

The task ahead is to elaborate the concept of wholeness in order to make us capable of overcoming the ideology of efficiency and prepare for a much healthier world, where we humans learn to make peace with the powers of Nature—in our minds and in our actions. And for this task, we all have to find our own Askja.

Bibliography

Berry, Wendell, *The Unsettling of America* (San Francisco: Sierra Club Books, 1977).

Hannesson, Pálmi, *Landið okkar* (Reykjavik: Bókaútgáfa Menningarsljóðs, 1957).

Heidegger, Martin, "Hegels Begriff der Erfahrung," in *Holzwege* (Frankfurt am Main: Klostermann, 1950), pp. 105–92.

Lovelock, James, *Gaia: A New Look at Life on Earth* (Oxford: Oxford University Press, 1979).

Lovelock, James, *The Revenge of Gaia* (New York: Basic Books, 2006).

Otto, Rudolph, *The Idea of the Holy* (Oxford: Oxford University Press, 1958).

Pascal, Blaise, *Pensées et opuscules* (Paris: Hachette, 1963)

Skúlason, Páll, *Meditation at the Edge of Askja* (Reykjavík: University of Iceland Press, 2004).

White, Michael and John Gribbin, *Einstein: A Life in Science* (New York: Dutton/ Penguin Books, 1994).

---

[26] Wendell Berry, *The Unsettling of America* (San Francisco: Sierra Club Books, 1977), p. 103.

PART 1
Restoring Place and Meaning

Mark H. Dixon

In *A Sand County Almanac*, Aldo Leopold's essential insight was that it was possible to see the natural environment as a system in which all the components (biotic and abiotic) were dependent upon each other on order to maintain an environment that was stable and so able to respond to climatological, geological and other changes. No less than other creatures, human beings are components within this system. In other words, natural environments are not abstract entities. Rather these are complex systems that humans participate in and influence through their decisions and actions. In turn natural environments influence both their human and non-human inhabitants as well.

Places are not static entities however, and problems arise when components within the system experience catastrophic change. Such changes ripple through and threaten the entire system. In the past these catastrophic events were themselves due to natural processes—earthquakes, tornadoes, hurricanes and so on.

Since the industrial revolution it has become more and more obvious that human beings are able to precipitate discernible and measurable environmental change on the same scale as these natural disasters. Though these often occur on a much slower time scale, these human-wrought changes are no less catastrophic.

This section endeavors to investigate how human beings can, and should, react to changes in place, in particular changes that are seen as negative or even catastrophic to the meaning and value that we assign to places. What we seldom realize though is that these changes do more than alter places (sometimes past all recognition), they also alter the manner in which we interact with these places—they undermine both meaning and value.

To understand this dynamic, this section examines environmental restoration. But just as environments are not abstract entities, so the human desire to alter places is something that happens through concrete and individual decisions and actions. Thus the authors in this section are interested in how restoration impacts particular places and the individuals that restore them. The following chapters include case studies of place, and how humans approach and seek to restore its moral and spiritual value.

Through these case studies and more reflective pieces this section explores the adverse impacts that human actions have had on natural and built environments and what our obligations are to correct these problems. All the chapters presuppose that such obligations do exist to restore human-wrought environmental damage. The questions that these chapters focus on probe other, perhaps even more basic questions: how do we restore environmental damage so that we restore meaning and value, as well as the environment? What attitudes ought to guide the restoration process? What technologies ought we to use in the restoration process?

The first two chapters represent more reflective pieces that focus on what ecological restoration is about and what role human beings ought to have in that process. What follows are chapters that deal with specific places and the particular problems that their restorations pose—the places in question include wilderness, urban areas and even human-built places. The differences in these places force us to consider restoration in many of its dimensions.

This section begins with "Restoration in Space and Place," in which William Jordan III, one of the founders of the discipline, reflects on ecological restoration as a holistic process—that is as an attempt to restore all an ecosystem's aspects. Jordan notes that what motivates such *ecocentric* restoration is a concern about the ecosystem in its own right, rather than more narrow human interests. The dilemma this creates though is that to maintain ecosystems in what approximates their "original" state requires that we compromise its "otherness" or "wildness" to some degree.

Todd LeVasseur's "Shame, Ritual and Beauty: Technologies of Encountering the Other—Past, Present and Future," builds upon Jordan's reflections, and indeed references Jordan's earlier *The Sunflower Forest*. LeVasseur is a participant with Jordan in The Values Project Study Group. What LeVasseur focuses on is the role that human participation has in the restoration process. What is it about our relation to the environment that inspires the desire to restore environmental damage and within what social context should such restoration occur? As the title indicates, prominent in LeVasseur's responses are the concepts "shame" and "ritual." It is insufficient to restore environmental damage without the recognition that the damage was ours. Genuine restoration requires us to acknowledge our role and to build into the restoration process rituals that allow both human beings and the environment to heal.

In Daniel Spencer's chapter, "Recreating [in] Eden: Ethical Issues in Restoration in Wilderness," we confront the specific problems that emerge once we acknowledge our ethical responsibilities to correct human damage to the natural environment. Spencer, a religious ethicist, asks: what happens when environmental restoration requires that we have to eliminate non-native species in order to re-introduce native ones? What Spencer attempts to articulate in the chapter is a decision-making model that would allow us to determine the most efficacious and well as ethical responses to such situations.

In Mélanie Walton's chapter, "Re-creation: Phenomenology and Guerrilla Gardening," we leave the lakes and deserts and move to an urban environment. Walton focuses on the potential that abandoned lots, medians and other forgotten areas represent in the attempt to (re-)introduce nature into the urban landscape. In Walton's opinion, it is the "guerilla gardener's" role to precipitate this process.

Biblical scholar Sampson M. Nwaomah's chapter, "Eschatology of Environmental Bliss in Romans 8: 18–22 and the Imperative of Present Environmental Sustainability from a Nigerian Perspective," considers what has become an all too common problem—the need to balance conservation against economic progress. Nwaomah's approach to this dilemma is to consider certain Biblical passages, in particular Romans 8:18–22, in order to determine what moral guidance is implicit within these passages as concerns the need to live in a sustainable manner.

In "Resurrecting Spirit: Dresden's Frauenkirche and the Bamiyan Buddhas" philosopher James Janowski focuses on questions about the obligation to remediate damage to the human built environment rather than the natural environment. Is there an obligation to correct damage that is due, for example, to wanton destruction (the Bamiyan Buddhas) or war (Dresden's Frauenkirche)? Janowski considers the extremes—to leave the sites as they are so that their absence serves as a testament to their loss, and to construct exact replicas with materials as similar to the original as possible. Neither option, Janowski argues, is acceptable, since neither will restore meaning or value. The response that Janowski advocates is *reconstruction*, i.e. to salvage as much original material as remains to rebuild the structures and to use similar materials where no original materials are available.

The section ends with a chapter that brings us full circle with a return to the natural environment, A. James Wohlpart's "Chanting the Birds Home: Restoring the Spirit, Restoring the Land." In this chapter Wohlpart, along with his students, take a trip down the Kissimmee River, which is under restoration to remediate the damage that was done during the river's channelization in the 1960s. The restoration is at different stages at different places along the river. Thus the trip down the river moves through "a remnant run cut off from its traditional flow, the channelized canal, and the restored river." As the group passes through each area, Wohlpart and students reflect upon the river's condition and its impact on the plant and animal species that can be seen in each section.

Chapter 2

Restoration in Space and Place

William R. Jordan III

We felt very nice and snug, the more so since it was so chilly out of doors; indeed out of bed-clothes too, seeing that there was no fire in the room. The more so, I say, because truly to enjoy bodily warmth, some small part of you must be cold, for there is no quality in this world that is not what it is merely by contrast. Nothing exists in itself.[1]

So writes Ishmael, reflecting on the experience of lying in bed in a chilly room at the beginning of the adventure he recounts in *Moby-Dick*. "If you flatter yourself," he continues, "that you are all over comfortable, and have been so a long time, then you cannot be said to be comfortable any more. But if ... the tip of your nose or the crown of your head be slightly chilled, why then, indeed ... you feel most delightfully and unmistakably warm."

It may be that this reflection on comfort—that too much of it results in a kind of sensual or experiential monotony that cancels it out—applies to the experience of place as well. Being too much "at home" in a place entails a danger not only of provincialism but of boredom, and not only boredom but loss of identity. Consider that if you know who you are by knowing who and what you are *not*, then you need the experience—perhaps troubling experience—of some other to achieve that.

Place, then, may be comfort, but place alone, an environment made familiar by long experience and a rich accumulation of memories, associations and meanings, may be an airless, claustrophobic and oppressive experience without at least a "tip of the nose" experience of cold and unfamiliar space to give it context: this familiar thing is not that other, strange thing.

How, then, can we bring the experience of space and the other into the gardened environment of the familiar? And what role might the practice of ecological restoration play in that?

During the past couple of decades, as land managers have discovered restoration and come to realize its value as a conservation strategy they have also come to appreciate its value as a way of re-entering and re-inhabiting classic ecosystems or landscapes such as the tallgrass prairie of the American Midwest that often serve as models for restoration efforts. This is a remarkable discovery. It makes it possible to bring back an ecosystem such as the tallgrass prairie. Prairies,

---

[1] Herman Melville, *Moby-Dick* (New York: Norton & Co., 1967), p. 55.

after all, once supported and were shaped by a human culture very different from the contemporary culture that now prevails and shapes the ecology of the region. This gives a future to an ecosystem that may be economically—and even ecologically—obsolete. Indeed, the prairie is strange, and so is felt strongly *as* other, at least partly because it *is* obsolete, is no longer the ecosystem of choice for the local economy, and is in ecological tension with its altered surroundings. It also resolves the dilemma of the human use of "natural"—that is, the historic— ecosystems that has bedeviled environmentalists since the time of John Muir: how to "use" such an ecological system without using it up, or altering its character.

These are important discoveries. But there is more. During the past few years I have been working with George Lubick, an environmental historian at Northern Arizona University, on a history of ecological restoration. Of course, ecological restoration is a very broad term, and can refer to anything that involves an attempt to bring back any aspect of an ecological system—its productivity, for example, or some feature that makes it valuable as a habitat for a particular species. That kind of land management—or restoration—has a long history. Historian Stephen Pyne notes, for example, that the use of fire to manage the environment—arguably a version of restoration—predates the emergence of our own species.[2]

George and I were interested, however, in what seemed to us a distinctive kind of restoration—that is, the attempt to restore not just selected features of an ecological system, but *all* of them. This is a kind of land management that has a much more recent history, dating back, as far as we can tell, only to the early decades of the past century, or perhaps, stretching the point a bit, the last couple of decades of the nineteenth century. Projects aiming to restore swatches of the tallgrass prairies of the American Midwest provide what many regard as the classic examples of this version of restoration, but projects with the same aim are now being carried out on behalf of a wide range of ecosystems, as readers of this book are well aware.

This business of restoring the whole thing, including all its "parts" and processes, is not necessarily practical. In fact, it often entails some sacrifice of immediate human interests, and that struck George and me as interesting—and as important.

Here is a form of land management that, while it certainly offers many benefits for the people involved, is not, in the last analysis, defined by their interests. It is true that in many instances restoring a "natural"—more accurately an historic, or merely old—ecosystem enhances the beauty of a place or increases its value as a resource, or repository of natural capital. In many cases, however, these utilitarian benefits do not justify the effort and expense of attempting to restore *all* of the species and processes that characterized the model ecosystem. Indeed, the claim that "all the parts" are necessary for an ecosystem to function properly begs the

---

[2]   Stephen J. Pyne, *Vestal Fire: An Environmental History, Told through Fire, of Europe and Europe's Encounter with the World* (Seattle: University of Washington Press, 1997), pp. 27ff.

question of what "properly" might mean, and is in any case a claim that, handy as it is for constructing arguments for keeping all the parts, ecologists no longer endorse.

There is, in other words, an impractical, altruistic element at play here. This is not—or not only—about maintaining human habitat, or about what ecologist Fikret Berkes calls the motive of "livelihood" that typically underlies the land-management practices of traditional, place-based peoples.[3] It is not about our interests, but is about self-consciously setting our interests aside, as necessary, in deference to the old ecosystem, to nature-as-given. The distinction is important— as important as the distinction between altruism and self-interest that, elusive as it is, nevertheless lies at the heart of human moral life.

To distinguish this from other, self-interested, forms of land management, George and I have termed this *ecocentric restoration*—that is, restoration of the whole ecosystem for, in the last analysis, its own sake.

Consider what is going on here. Classic examples of the kind of restoration we are talking about, examples that most practitioners and critics would agree are representative of the type, entail an effort to return an ecosystem or a landscape to the condition it was in before "we" arrived on the scene. "We," of course, is a technical term here that, like the designation of a wild card, can mean anything the dealer chooses. In the United States and Australia, where this form of land management was invented, in part as a response of newcomers to the ecological changes resulting from their presence, "we" has typically meant European immigrants or, if you prefer, invaders, and the horizon defining the ecosystems chosen as models for restoration projects has been the local or regional version of "1492." But "we" can also mean "our family" or just "me" or even human beings. There is no reason why a restorationist might not aspire to return an ecosystem to the condition it was in prior to the arrival of humans, and in fact this is being done on a sizeable scale at the Kaori Sanctuary in New Zealand.[4]

However "we" is defined for the purposes of a restoration project, the project winds up being committed to the re-creation of an ecosystem or landscape that is not about "us." It was here before "we" got here. In its original manifestation it owed nothing to "us." And the attempt to restore it, to summon it back into the landscape, amounts to an attempt on the part of the restorationist to disappear ecologically—to behave in such a way (a very odd way, actually) that the ecosystem can resume behaving *as if* "we" had never arrived.

It is not, in other words, about "us" except in the special sense that it is about the "not us" against which we define ourselves in ecological terms—an essential

3   Fikret Berkes, *Sacred Ecology* (Second Edition) (New York: Taylor & Francis, 2008).

4   www.sanctuary.org.nz/Site/Conservation_and_Research/Restoration/The_fence. aspx. For a discussion of ideas about restoration of pre-human ecosystems in New Zealand developed by paleontologist Matt McGlone, see Lesley Head, *Cultural Landscapes and Environmental Change* (London: Arnold, 2000), pp. 103–5.

step in achieving citizenship in a land community in which "we" can fashion some kind of comity with all these others.

By the same token it is "space," a strangely self-conscious re-creation of the "howling wilderness" the Puritans experienced in Massachusetts. This was home—place—to hundreds of thousands of people. But to the newcomers it was strange, fascinating and forbidding—the space against which they began to define themselves, and within which they began to create places of their own.

Obviously, re-creating the old "wilderness" or "space" is an odd and impractical thing to do. Though this effort certainly provides insights into practical matters such as sustainable agriculture and other forms of land-use, it does not offer a paradigm for sustainable land-use in general. They may—or may not—be economically useful—competitive on economic terms with frankly working landscapes such as pastures, cornfields and ornamental gardens. And they are unlikely to be self-sustaining. Indeed, in most cases they will require continual restorative management if they are to persist as reasonable likenesses of the old ecosystems they are intended to represent in the context of the altered landscapes in which they find themselves resurrected.

They may not—should not; indeed, cannot—be the static, "snapshot" representations of old ecosystems that restorationists rightly deplore. But they will be museums of a sort, relics and imperfect representations of ecosystems that restorationists rightly disparage.

Certainly, in some cases, they may, unlike the objects in a museum, be both useful and ecologically viable; that is, they may serve as human habitat, as providers of natural resources, as the restored salmon habitat in the Pacific Northwest, or the restored pine savannas of the Southwest do. But even—indeed especially—when this is not the case it serves as an emblem of otherness, of strangeness. It represents the "not us" that was here when "we" got here—the spaces within which we have created places, and which provide the context within which we create meaning, and which provide the "tip of the nose" contact with the other that Melville's figure of a man discomforted by too much comfort suggests we need.

It is for just this reason that Melville's contemporary, Henry Thoreau, suggested that "every town should have a park, or rather a primitive forest, of five hundred or a thousand acres, where stick should never be cut for fuel,"[5] an early instance of the appeal for wilderness preservation that has been an important element in environmental thinking for nearly two centuries.

This can be achieved, I should note, merely by "preserving" a landscape—that is, by fencing it off and leaving it alone, as Thoreau suggests—but only at the expense of the ecological community, which, isolated in the context of a changing landscape, will drift ecologically, both losing and gaining species as it gradually changes into something else.

---

[5]   Henry David Thoreau, Journal XII, 387, in *The Journal of Henry D. Thoreau*, Bradford Torrey and Francis H. Allen (eds) (New York: Dover, 1962), Vol. 2, p. 1529.

In order to maintain an ecosystem in at least an approximation of its "original" condition under altered conditions it is necessary to compensate continually for the novel influences on it—a management protocol that amounts to an ongoing program of ecocentric restoration.

This may compromise the otherness, or wildness of the system. But that is the price we have to pay for keeping the ecosystem—and many of the species that compose and inhabit it—and ensuring their survival and well-being into the future.

It is the cost, we might say, of the change we bring about. Or to put it another way, a valuable occasion for reflecting on those changes and their ecological consequences.

And there are precedents for this. The clearest one I know is the institution of the sacred groves. Common in the ancient Mediterranean, and still vital elements in the landscapes of India and parts of Africa today, these are areas set aside in settled, working landscapes as a tribute to the gods of the place—to, we may say, creation in its aspect as given, and owing nothing to us.[6]

Another is the institution of the Sabbath, which enjoins the setting aside of work, not as vacation, a "break," or just kicking back, but as a deliberate, disciplined gesture of respect for the given. Theologian Norman Wirzba argues that in the Judeo-Christian tradition, creation reaches its epitome and fulfillment in this "rest ... of God" in which the creator contemplates the creation.[7]

And to this the philosopher Josef Pieper adds the idea that this "leisure" is not only the basis for culture, it must be centered on public ritual—certainly a profound challenge to what Pieper calls a culture of "total work" in which only work is valued and all value is the result of human work.[8]

If the claims of the Sabbath are real, then every place needs this space around it, this sacred grove, or actively maintained Sabbath of the land, not only to conserve species, but also to lend the place context and meaning.

And if Professor Pieper is right then perhaps the wisest thing to do is to make the creation and maintenance of such a space what it has always been for those who maintained the sacred groves—that is, not work but play and an occasion for contemplation, celebration and festival.

---

[6]   M. Jha, H. Vardhan, S. Chaterjee, K. Kumar and R. R. K. Sastry, "Status of Orans (Sacred Groves)," in P. S. Ramakrishnan, K. G. Saxena and U. M. Chandrashekara (eds), *Conserving the Sacred for Biodiversity Management* (Enfield, NH: Scientific Publishers, 1998). Prof. Berkes notes that the sacred groves are an exception to his observation that land-management practices of pre-modern societies are characteristically "livelihood" oriented.

[7]   Norman Wirzba, *The Paradise of God: Renewing Religion in an Ecological Age* (Oxford: Oxford University Press, 2003), pp. 34ff.

[8]   Josef Pieper, *Leisure the Basis of Culture* (San Francisco: Ignatius Press, 2009). Pieper makes the point I make here very concisely in his author's preface, pp. 15–16, but the idea is the key point of the entire essay.

Bibliography

Berkes, Fikret, *Sacred Ecology*, Second Edition (New York: Taylor & Francis, 2008).

Head, Lesley, *Cultural Landscapes and Environmental Change* (London: Arnold, 2000).

Jha, M., H. Vardhan, S. Chaterjee, K. Kumar and R. R. K. Sastry, "Status of Orans (Sacred Groves)," in Ramakrishnan, P. S., K. G. Saxena and U. M. Chandrashekara (eds), *Conserving the Sacred for Biodiversity Management* (Enfield, NH: Scientific Publishers, 1998).

Melville, Herman, *Moby-Dick* (New York: Norton & Co., 1967).

Pieper, Josef, *Leisure the Basis of Culture* (San Francisco: Ignatius Press, 2009).

Pyne, Stephen J., *Vestal Fire: An Environmental History, Told through Fire, of Europe and Europe's Encounter with the World* (Seattle: University of Washington Press, 1997).

Thoreau, Henry David, *The Journal of Henry D. Thoreau*, Vol. 2, Bradford Torrey and Francis H. Allen (eds) (New York: Dover, 1962).

Wirzba, Norman, *The Paradise of God: Renewing Religion in an Ecological Age* (Oxford: Oxford University Press, 2003).

Chapter 3

Shame, Ritual and Beauty: Technologies of Encountering the Other— Past, Present and Future[1]

Todd LeVasseur

This chapter is part of a larger dialogue in which I am a participant, organized and spearheaded by William (Bill) Jordan III, a leader of the ecological restoration movement that has taken shape over the past 25 years. The goal of the dialogue is to add to and further develop the ideas presented in his 2003 book *The Sunflower Forest*.[2] In this book, Jordan builds upon a theory of value developed by the literary critic and historian Frederick Turner. Jordan does this by applying Turner's theory to the human–nature relationship that has become the focus of environmental ethics (and the field of religion and nature, out of which I operate). Writing as environmental historian, environmental ethicist, biologist and observer of the emerging culture of restoration, Jordan suggests that Turner's theory of value creation applies to the human–nature relationship; furthermore, he argues that the act of ecological restoration, when ritualized, provides a valuable context in which to explore and negotiate the relationships between humans and the "natural" environment. This, he suggests, includes the ecological and economic dimensions of the nature/culture relationship. It also includes valuable opportunities for the development of ritual, which is a key element in Turner's theory of value. Such eco-ritual, described throughout this chapter, prepares and germinates a seed of higher value creation that can allow humans to achieve a deeper relationship

[1] I want to extend my thanks and gratitude to William Jordan, III. These thanks are not only for his bringing our group together and serving as personal and intellectual inspiration, but also for his deft work in editing and commenting upon the original draft of this chapter. Without his patient reading and many thoughtful comments, this chapter would not be as clear in its summaries of his work, Rappaport's and Turner's work, and the work of The Values Project Study Group.

[2] See The Values Project Study Group's forthcoming manifesto, "An End to Innocence: A Model for an Environmentalism for Grownups," currently in rough draft form and to be published in 2011. See also William Jordan III, *The Sunflower Forest: Ecological Restoration and the New Communion with Nature* (Berkeley: University of California Press, 2003).

with nature. This relationship is one based on engaged, mindful, performative interaction, mediated by ritual, and that centers on the

> emotionally demanding conception of relationship [and] the idea that there is an
> essential, unbreakable link between the experience of transcendent values such
> as beauty, meaning, and community and the experience of shame that arises
> from our awareness of the world and the sense of limitation and difference this
> awareness entails. Shame here, it is important to emphasize, is distinct from guilt.
> It is not the response of the conscience of what we *do*, but of our consciousness
> of what we *are* … Shame … is the emotional register of our natural, radical,
> existential dependency and a debt for which we are not responsible and which
> we cannot pay.[3]

In another setting, Jordan writes that his "notion of shame and its relationship to performance and value [is that] shame [is] an emotional response to shortcoming … [and] it is also [an important and under recognized] pathway to value."[4] Because this is a universal aspect of experience—all of us must premeditate upon what we are to cull, kill and consume from Others that gift us life; and because at our core we know and/or feel that we cannot repay these gifts that entail harm, suffering and death of non-human Others (and the larger ecosystems in which all life forms are embedded)—we experience shame.

Jordan couples this hypothetical ontology of shame with the insights of David McCloskey, who writes that the three modern myths of community are that communities (especially human–human communities, but also human–nature communities)[5] are harmonious, homogenous and solidary.[6] Jordan labels this view of community, and the animating view of contemporary environmentalism that holds to these views of community, as "sentimental." This hopeful, naïve view of our community relationships does not demand of us that we deal with the untoward, problematic aspects of our existence and relationship with the planet,

---

[3]   Jordan, *Sunflower*, pp. 46–7.

[4]   Jordan, email correspondence with VPR Group on August 21, 2007.

[5]   David McCloskey, "What Community is Not: Three Myths about Community" (unpublished manuscript, Seattle University, 1997). I recognize that language is problematic and that "humans" are actually primate animals who are wholly part of nature. It could be argued that this distancing language is part of our embedded vernacular because it precisely hides the existential challenge of which Jordan writes. In part, our language creates a dualism (of both human/nature and nature/culture) based on a false consciousness embedded in our language that implies we are subjects acting in a world of objects. Therefore, the current anthropogenic extinction crisis is possibly occurring in part because we see ourselves as "humans," and thus separate and superior from the rest of creation. For a more in-depth look at human exceptionalism in general, see Anna Peterson, *Being Human: Ethics, Environment, and Our Place in the World* (Berkeley: University of California Press, 2001).

[6]   Jordan, *Sunflower*, p. 44.

including especially our shameful dependency on processes of life and death that are greater than us. For Jordan, building on the work of Turner, the earth offers us gifts that we by definition cannot repay—thus, we experience shame. And while Turner and others present evidence that this experience is universal, Jordan writes that:

> The idea that there is an essential link between an ultimate value such as community and negative feelings such as fear, horror, or shame is foreign to environmentalism, as it is to modern thinking generally. It is this failure to confront the trouble and shame of creation that has limited what environmentalism has been able to accomplish.[7]

In essence, he argues that environmentalism has failed because it does not address this deeper human experience and thus misses an opportunity to create new values related to our immediate environments. Thus, the sentimental environmentalism that has developed in the West does not provide us with psychologically and spiritually productive ways of dealing with what he calls shame. In short, it does not recognize that shame exists and therefore can offer to humans no wisdom on how to process and effectively engage it. For Jordan, one way to deal with this existential shame is through engaging our emotions by conducting ritual. Jordan argues that via ritual shame can be named, unmasked and engaged, and higher values of beauty and respect can emerge in our relations with the Other/these Others who give us life.[8] Such existential shame is foundational to being a culture-bound biological being and must be engaged if we are to (hopefully) move out of our ecocrisis and the actions and worldviews that in large part have spawned it.[9]

Given the above, I argue that in order to access this neglected aspect of our existence, and thus for environmentalism to be both effective and affective, humans should in some way utilize the technology of ritual as a valuable way of accessing and creating value. I argue that humans can enter into relationship with community and one effective way is by engaging in performative, ritualized ecological restoration. For Jordan, ritualized restoration is a "process

---

[7] *Ibid.*, p. 46.

[8] My use of the terminology of naming, unmasking and engaging is inspired by the work of Walter Wink (New York, 1999).

[9] Here when I write "we" I refer to those whom Daniel Quinn calls "Takers," and thus not to all humans (*Ishmael* (New York: Bantam/Turner, 1992); *The Story of B* (New York: Bantam/Turner, 1996)). Quinn posits that humans, East and West, despoil their environments. See Yi-Fu Tuan ("Discrepancies Between Environmental Attitude and Behaviour: Examples from Europe and China," in David Spring and Eileen Spring (eds), *Ecology and Religion in History* (New York: Harper and Row, 1974)) for an example of scholarship that explores China's various historical and deleterious relationships with its landbases. It must be pointed out that, as Jordan argues, some cultures do recognize this shame of dependency and engage it with ritual to access transcendent values.

of community building, or entry into community"[10] that entails four stages: (1) achieving awareness of the other (in this case, a landscape to be restored); (2) developing an ecological/economic relationship with the other; (3) an exchange of gifts with the other (based on the recognition that all ecological relationships are economic in basis, where an inequitable exchange of goods and services takes place and dramatizes the shameful inequity of any exchange of gifts); and (4) resolving the ambiguity inherent in this exchange by ritual and myth, which allow us to encounter our shame and move through it to even more transcendent values, including communion and community.[11] Drawing on Turner's work and the work of anthropologists Victor Turner and Roy Rappaport, Jordan argues that, "ritual ... is fundamentally creative ... Most fundamentally, [ritual] is the means by which humans generate, recreate, and renew transcendent values such as community, meaning, beauty, love, and the sacred, on which both ethics and morality depend."[12] See here that for Jordan and our group, morality and ethics in large part results from enacted ritual in a landscape.

The Technology of Ritual

With this quick overview in place, I will now further explore the technology of ritual and explore the role it may play in helping us move beyond shame and into an effective, affective and restorative environmentalism. To begin, I will build upon Jordan's work. I start with a quote from another Frederick Turner (i.e. not the one Jordan draws upon), who in the book *Beyond Geography* writes that, "No culture desires to be wholly at the mercy of nature."[13] This is hinting at what Jordan calls shame: the dialectic between recognizing our limits (both individual and collective) within and ultimate dependency upon nature—being at nature's mercy—and wanting to escape this predicament, yet being unable to, and how value emerges out of this dialectic. I argue that ritually based restoration ecology is a technology humans can develop and utilize as an attempt to engage the emotions that arise from experiencing our inequitable economic dependency upon nature. Furthermore, by using ritual to creatively and performatively engage our feelings of shame, we can generate community and provide access to the values on which an effective environmental ethics may be based.

This same Turner writes about a spiritual vision that came to him in the 1970s as he was visiting the Black Hills region of South Dakota. Turner's vision showed him that he was "estranged" from the mythical and spiritual history of the American landscape. Turner writes about this vision that, "this was ... my own

---

[10]   Jordan, *Sunflower*, p. 51.

[11]   *Ibid.*, p. 51.

[12]   *Ibid.*, p. 5.

[13]   Frederick Turner, *Beyond Geography: The Western Spirit Against the Wilderness* (New York: Viking, 1980), p. 21.

ignorance of America, an ignorance amounting to an estrangement from the land, [and] was in some real measure that of Americans generally."[14] Keep in mind that the estranged Americans whom Turner is talking about in this passage are the progeny of European colonists. Wendell Berry and Wes Jackson both echo this same sentiment, arguing that descendants of European colonists and immigrants suffer from a spiritual estrangement from the land. These cultural critics are saying we have not, to use Wes Jackson's phrase, "become native to this place," and for all three, this results in a profound spiritual disconnection.[15] The ritual-based prescription offered in *The Sunflower Forest* and in the subsequent generation-of-value project we are working on is one potential antidote to this spiritual eco-anomie. Equally, if we are to take Turner, Jackson, Berry and Jordan seriously, then we need to look at spiritual, or, more broadly, religious approaches to recreating, replacing and restoring our collective Euro-influenced (dis)connection/s to the land and the various landscapes of America.

I further my argument by briefly turning to the work of the same Frederick Turner who Jordan engages and share here Turner's thoughts regarding religion and ritual. What Turner says provides an entry point for looking at ritual in general, and for my subsequent discussion about the anthropologists Victor Turner (Frederick Turner's father) and Roy Rappaport, in particular. The younger Turner, Frederick, in a claim that is somewhat problematic because it is totalizing—yet which is nonetheless provocative—states that, "We are a religious species ... we repress the religious drive at our own peril."[16] He describes traditional institutional religion as "imaginative psychic technologies of great age and effectiveness" and asserts that "where a religion really stands or falls is in its ritual."[17] For Turner, ritual carries "the inherited traces of our early evolution—the great psychic technologies of mythic storytelling, chant, sacrifice, body decoration, music, dance, the fresco."[18] Furthermore, he maintains that if these psychic technologies that have impacted the evolution of our brains, bodies and societies are present in ritual, then a religious theology and ethics will be generated, for they will be encoded within these technologies.[19] This brief précis on Turner's thoughts about

---

[14] *Ibid.*, p. 5.

[15] Or as sociologist Robert Bellah writes ("Meaning and Modernity: American and the World," in Richard Madsen, William M. Sullivan, Ann Swidler and Steven M. Tipton (eds), *Meaning and Modernity: Religion, Polity, and Self* (Berkeley: University of California Press, 2001), pp. 255–76), Americans harbor an extreme "ontological individualism" that results in disembeddedness from other humans, and also disembeddedness from the land and from relations of human–non-human (my interpretation).

[16] Frederick Turner, *Rebirth of Value: Meditations on Beauty, Ecology, Religion, and Education* (New York: SUNY Press, 1991), p. 83.

[17] *Ibid.*, pp. 84–5.

[18] *Ibid.*, p. 86.

[19] While Turner's work goes back to the late 1950s, it is supported by recent research in neurophysiology. For example, the anthropologists John Baker and Michael

ritual provide an entry into discussing Roy Rappaport's theories about these same ritual and psychic technologies.

Rappaport's early work focuses on the rituals the Maring people of New Guinea employ in their environment, especially regarding the hunting of pigs. Rappaport utilizes a structural-functionalist anthropological approach in this early fieldwork and hypothesizes that rituals have empirical effects on the environment. In effect, rituals are part of a "behavioral repertoire employed by an aggregate of organisms in adjusting to its environment."[20] In other words, in this earlier work, Rappaport claims that ritual helps human groups function as an organism in an ecosystem. He further maintains that ritual regulates relationships in complex systems so that group behavior falls within a reference value or goal range within the system.[21]

On one level, we can see how eco-restoration ritual falls into Rappaport's definition of ritual. Ritual eco-restoration enables humans, acting in aggregate as an organism embedded within an environment, to culturally and biologically interact with and restore ecosystems. Restoration becomes a form of economic engagement with the environment that provides a context for the invention of rituals that enhance our relationship with the environment, while also providing means of dealing with problematic aspects of that relationship. Furthermore, our group stresses that restoration is only one of many economic contexts where ritual can be created (see "Constructing a Hypothetical Eco-Ritual," below).

It must be noted that Rappaport points out that the more economically and thus socially differentiated a society becomes (especially when coupled with an increase in population), then the more economic rationality becomes separated from ecology so that a culture can develop to be at extreme odds with nature. This is something to keep in mind, as many humans today (including the approximately 50 percent dwelling in urban areas) live within a culture that privileges economic

---

Winkelman indirectly support Turner's claim, writing that, "Religious experiences are integrative, enlightening, and meaningful because they stimulate the centers of the brain where information integration and a sense of certainty about the truth of what we know are produced" (John Baker and Michael Winkelman (eds), *Supernatural as Natural: A Biocultural Approach to Religion* (Upper Saddle River: Pearson Prentice Hall, 2010), p. 70). In other words, and as Roy Rappaport argues, emotionally charged mores and values can be transmitted during ritual, including information about our local environments.

[20]   Roy Rappaport, *Ecology, Meaning, and Religion* (Berkeley: North Atlantic Books, 1979), p. 28.

[21]   As Ronald Grimes points out, "The import of the ethnographic testimony is that ritual participants the world over believe their rites enable them to live in synchrony with the natural world, especially animals and plants, sometimes even rocks, mountains, bodies of water, and specific places. This ritual sensibility is not characteristic of every ritual system, but it is typical of many of them, especially the local, smaller-scale ones" (Ronald Grimes, "Ritual," in Bron Taylor (ed.), *The Encyclopedia of Religion and Nature* (New York: Continuum, 2005), p. 1386). Our group argues that this synchrony (we are ambivalent about such a state becoming "obtainable") actually entails working with the emotions and feelings of shame.

rationality and mechanistic science. Both of these worldviews tend to divorce our emotions from recognizing our inequitable relations with the landbases upon which we depend for existence. They both therefore neglect to offer adequate contexts and occasions for various rituals that can possibly lead to the creation of community and a valuing of our local environments.

Rappaport built upon this earlier work by turning his attention to how ritual can be used by humans to help foster sustainable ecological relationships. His work suggests that ritual plays a role in developing values and ethics that can lead to living within environmental limits. This concern with ecological sustainability is central to his capstone book, *Ritual and Religion in the Making of Humanity.* Rappaport begins this book by highlighting the importance of verbal language. The evolutionary development of language led to its advent and subsequent use as a conceptual tool that allowed humans to create religion and to formulate symbolic concepts about their environments. Rappaport points out that language allows for lying and false representation; as a result, ritual developed to help generate group cohesion, sociability and trust in order to offset lying and falsity. He defines ritual as "the performance of more or less invariant sequences of formal acts and utterances not entirely encoded by the performers."[22] Such performance of invariant sequences of formal acts and utterances leads to convention, social contract, morality, construction of time and space, awareness of the divine, and "the construction of orders of meaning transcending the semantic."[23] The constructs that result from ritual are symbiotically produced with ritual's form, without which ritual could not exist. For Rappaport, the four main defining elements of ritual are: encoding by other than the performers of the ritual (which includes coding by tradition and/or an appeal to an Ultimate), formality, invariance and performance. Thus, the form and content of ritual create certain meanings not able to be created by any other human activity—hence the ubiquity of ritual across the globe, making it the "social act basic to humanity."[24]

Of import for Rappaport (and my argument about ritual eco-restoration) is his claim that ritual carries social authority and generates both obligation and duty, and thus, morality. To the extent this is true, ritual carries within it both self-referential and canonical indexical messages so that the simple act of performing a ritual makes the performer subject to the discipline of the ritual's invariant order and meaning system. Such discipline entails a deference to the Other that is crucial to the development of a relationship with this Other. Ritual also provides a buttress against forces of cultural and ecological dissolution, helping keep the world coherent for its practitioners. This means that a ritual designed to restore the environment not only automatically brings the performer into its symbolic, ethical and performative meaning system but also serves both to maintain that

---

[22] Roy Rappaport, *Ritual and Religion in the Making of Humanity* (New York: Cambridge University Press, 1999), p. 24.

[23] *Ibid.*, p. 27.

[24] *Ibid.*, p. 31.

system and to negotiate its adaptation to changing conditions. Such performance also provides a coherent, place-based ecological worldview to those participating in an eco-ritual. Such a coherent worldview can help offset the centrifugal forces of omnipresent consumer advertising and an economic system that externalizes and denies our shameful dependency upon natural systems.

For Rappaport, ritual time helps to generate the feeling of *communitas*. This is because ritual is involved in organic, social and cosmic time, and, as he writes, "the rhythms of the [ritual] order reach in two directions at once—into each participant's physiology on the one hand and outward to encompass all of the participants on the other."[25] I add that this outward reaching of ritual can and indeed does include nature as a participant, and thus brings nature into contact with each participant's physiology, especially in eco-restoration rituals. Or, as David Crouch writes, "performance enables the movement away from (rather than necessarily in opposition to) prevailing/dominant versions of what nature is and in what ways it may matter."[26] As such, some roles of eco-restoration ritual are:

- It has the potential to engage the bodies, emotions, minds and morals of all the participants in the ritual.
- It engages the non-human world with which the ritual is in dialogue and engaged relationship.
- It builds morality (Crouch's "what matters") into these varied engagements.
- It moves us away from both our dominant idea of humans being separate from nature and from our sentimental view of nature that denies shame.

Rappaport highlights another important aspect of ritual, in that ritual assumes a constant changelessness around which history unfolds. This is in part because of the role and function of what Rappaport terms Ultimate Sacred Postulates (USP). Ultimate Sacred Postulates are, in Rappaport's view, not just expressed by the ritual, but are actually constituted by the performance of liturgical ritual systems. This means they have an enduring public nature, so that participation in a ritual entails both acceptance of and obligation to USPs. In other words, the technology of ritual liturgically encodes and performatively enacts Ultimate Sacred Postulates that in turn shape and guide group morality over time. Therefore, the "liturgical order" of evolution's "recurrent changelessness" and the laws of ecology thus can join to become the organizing axis/USPs of human activities, as compared to our current linear and/or transcendent economic, ethical and religious views of history.

---

[25]   *Ibid.*, p. 228.

[26]   David Crouch, "Performances and Constitutions of Natures: A Consideration of the Performance of Lay Geographies," in Bronislaw Szerszynski et al. (eds), *Nature Performed: Environment, Culture and Performance* (Oxford: Blackwell, 2003), p. 27.

Ritual, for Rappaport, is an adaptational response to an environment that is both always unknowable and inherently meaningless.[27] The environment is given meaning by human constructs and ideas, with the ritual based "Cybernetics of the Holy" (reifying USPs by encoding them into religiously sanctioned ritual/s) being a core feature of humanity that gives meaning to human life. Ritual can also allow participants to move away from the conscious reasoning that tends to abstract humans out of their embedded evolutionary and ecological reality. Thus, participants in eco-ritual are offered an antidote to the eco-anomie about which Frederick Turner, Wendell Berry and Wes Jackson write. However, the sanctity of ritual can be revoked by the masses and by the "Generation of the Lie."[28] If this happens, the adaptive benefits of ritual can become maladaptive, especially if sanctification of USPs creates an us/them mentality (including a human/nature mentality). Or ritual can become maladaptive if there is an over-sanctification of highly specific sets of rules, which can lead to dissonance and withdrawal of support. In other words, like any technology, ritual, including eco-ritual, can fail in its performance aspect. It can also fail in its group morality aspect. If Rappaport is right, then creators and performers of eco-rituals need to be on guard against these possibilities of failure.[29]

Rappaport ends his book by advocating a post-modern science grounded in ecology as the new Logos—or concept of the Holy—around which society should organize itself. He claims that the evolution of secular knowledge based on positivist science has become the modern day USP under which humans operate; however, for Rappaport these secular facts contain no numinous qualities so that the Logos (ecology, the environment) is threatened.[30] Because sanctity has the

---

[27] Here "adaptive" is used in an evolutionary sense, rather than in the socially conserving function of ritual.

[28] Rappaport points out that, "Grammar makes it possible to conceive of alternative worlds" (Rappaport, *Ritual*, p. 17), some of which are false worlds purposely created for power and control. For example, oil companies discredit studies that suggest anthropogenic climate destabilization is rapidly occurring.

[29] Along similar lines, Ronald Grimes writes that, "Without a ritual-dramatic stage between the narrative experience and the ethical judgment we are extremely subject to self-deception concerning the degree to which we have embodied our ethics. Without a keenly developed ritual-dramatic sense, our narratives are at best intellectual ideals and at worst sources of heteronomously imposed, introjected images" (*Ritual Criticism: Case Studies in Its Practice, Essays in Its Theory* (Columbia: University of South Carolina Press, 1990), p. 165). This supports what I am arguing in this chapter: ritual encodes ethics and also generates the values on which any ethics is based. Without incorporating into our environmental ethics and practices this ancient technology that has influenced the evolution of our brains and bodies and environments, we are just speaking sentimental narratives that are not only devoid of any real power—power to change our actions, and power to encounter and deal with shame—but may be, as Grimes argues, dangerously self-deceiving.

[30] Some, like David Loy, argue that, in Rappaport's terms, neoliberal market economics has provided us with another set of USPs and that the values codified in these

potential to support social orders and increase the adaptiveness of social systems (it also has the potential to do the opposite), the USPs given in ritual to a community of practice are of vital evolutionary importance. This is why the technological cybernetics of ritual are needed: humans require beliefs and practices that can ground them in the larger biological world—beliefs that give both humans and the larger biological world meaning—so that humans can overcome the Lies of grammar and fictitious worlds.[31] In our case, a cultural Lie that ritual, especially eco-ritual, can overcome is the Lie that we are superior to nature. Another Lie eco-ritual can possibly overcome is our at-large cultural denial that being part of a community entails shame. The USPs of eco-rituals can potentially create an eco-morality that is able to ritually codify and performatively deal with shame and transcendent values and thus undergird within communities an engaged, effective environmental morality. I argue that eco-restoration, as a ritualized technology, can help generate localized communities and institutions that recognize shame and are able to work with this foundational ontology to generate more sustainable relations with local ecosystems.

Let me now briefly turn to the work of anthropologist Victor Turner and his functional understanding of ritual and of performative acts in general. Turner is most well known for his theories of liminality and social drama. He writes that, "both ritual and theatre crucially involve liminal events and processes."[32] In this respect, ritual eco-restoration is a liminal space where shame-based values and morality can be accessed, created and disseminated via USPs based on the Logos of ecology. Ritual eco-restoration also becomes a vehicle and context of meta-commentary regarding the deeper values that both constitute and result from our relations with the Other. This includes creating a context where the reality of shame (recall the inequality of economic exchanges in nature—the "debt" humans accrue and can never repay to the community of local landbases of which we are part) can be accessed, leading to the creation of values based upon community and ecological USPs.

Turner's writings also deal with how communities undergo a social drama. For Turner, a social drama is a structured movement through time that includes (1) a Breach (of, say, an Ultimate Sacred Postulate), (2) a Crisis this breach engenders, (3) a Redressive Action, often ritualized, to deal with the crisis (and for Turner, this redressive action is enacted either in law or religion) and (4) if the drama is

---

postulates are inimical to healthy biological and human communities ("The Religion of the Market," *Journal of the American Academy of Religion* 65 (1997): pp. 275–90).

[31]   See here especially the work of David Abram (*The Spell of the Sensuous: Perception and Language in a More-Than-Human World* (New York: Vintage, 1997)), who writes of the "lie" of written, phonetic language and how this distances us from the "more-than-human" world.

[32]   Victor Turner, "Are There Universals of Performance in Myth, Ritual, and Drama?" in Richard Schechner and William Appel (eds), *By Means of Performance: Intercultural Studies of Theatre and Ritual* (New York: Cambridge University Press, 1990), p. 8.

allowed to run its course, a final stage of the drama unfolds that leads to either community reintegration or community schism.

Turner's view of ritual and performance is both ambitious and optimistic. He writes that, "True theater [via performance creates a] sense of harmony with the universe ... and the whole planet is felt to be *communitas*. This shiver has to be won, achieved, though, to be a consummation, after working through a tangle of conflicts and disharmonies."[33] In other words, Turner is arguing that either judicial laws and/or performative rituals deal with social dramas and the ritual part of this process leads to *communitas*. I maintain that this consummation, and thus *communitas*, is by definition impossible due to our inadequacy and inability to fully reciprocate gifts (especially economic) to our larger environment upon which we depend for our existence. Yet, by a formalized eco-ritual, we nonetheless attempt to reach eco-*communitas* with nature and by attempting this we recognize our limitations (including that we are not sovereign beings, but are in active, continuous relationship with an Other that gifts us our survival). We recognize that the schism and break, the "eco-drama," if you will, and the shame that is generated by this will always be present. Eco-ritual can potentially lead to short-term re-integration, especially in smaller communities, but the schism can never be done away with. Nonetheless, by the technology of ritual we can attempt to constructively use the shame Jordan writes about; encounter it in a ritual setting; use it to lead to higher values and positive experiences; and create and codify Ultimate Sacred Postulates based upon an ethical Logos of ecology, sustainability and direct participation in local ecosystems. In this way, actions and ethics, or values and practice, may become mutually reinforcing, leading to value formation. These values underwrite a proclivity for more action, helping to create a liturgical cycle of eco-rituals (as seen in the next section of this chapter). Through this ritualized restoration, USPs can be created and internalized, helping to lead to engaged and effective environmental ethics and practices, open to any humans interested in participating in the eco-ritual/s.

Constructing a Hypothetical Eco-Ritual

Given the above exploration of ritual, the present section focuses on the construction of an eco-ritual. To begin, I once again return to the Frederick Turner whom Jordan engages. Turner writes of the restoration of a prairie that "the whole affair resembles a ritual, complete with ecologist-shamans, adepts, novices, ordeals, and mystical instruction. This procedure engenders an extraordinary familiarity with the land."[34] Along these same lines, the religious studies theorist J. Z. Smith writes that, "Ritual is, first and foremost, a mode of *paying attention*. It is a process for *marking interest* ... It is this characteristic, as well, that explains

---

[33] *Ibid.*, p. 13.
[34] Turner, *Rebirth*, p. 66.

the role of place as a fundamental component of ritual: place directs attention."[35] In these quotations we see the centrality of place, of nature, and of paying attention to nature and becoming familiar with place, in the construction and function of ritual.[36] Through ritual, place becomes part of the Logos; through ritual, our radical, embodied shameful dependency on place becomes recognized and engaged. Since Aldo Leopold, environmental ethics has been telling us we must pay attention to the land, that we must value it morally; yet, environmental ethicists have long neglected the role of ritual and the role it should (and must) play in this process. To more fully pay attention to the land, and thus value it, we must bring in ritual and its performance.

Let me now give a brief example of what an eco-ritual might possibly look like. For readers who want a specific example of a theoretical eco-restoration ritual, see Jordan's *Sunflower Forest*; I will instead describe an agrarian ritual that engages the goals of sustainable agriculture.[37] My own background—both personal and professional—investigates agrarian values and how religious communities in the United States are combining religious environmentalist values with agrarian concerns for land and society. I have also worked on certified organic farms in California, South Carolina, New Hampshire and Scotland. As such, I have spent many hours engaging with and working on the managed ecologies of farmlands: lands (and seeds, plants and animals) that we are postulating we are shamefully dependent upon for our sustenance and survival. My own experience and passion thus lead me to want to bring ritual into eco-agriculture.

For example, this move and need is hinted at in the following passage from Wes Jackson, Director of the Land Institute in Salinas, Kansas, who writes, "We need food and, given the current human population now dependent on till agriculture, we will need to continue to till the soil, even though such activity has historically and prehistorically undercut the very basis of our existence. We live in a fallen world ... we must do our best to prevent further deterioration of this fallen world; indeed ... we must restore it."[38] Not only must we restore it, but we must be active participants in this restoration: from creating the vision of what we are going to restore the land to, to how we are going to do this. As Jordan points out, eco-restoration is about deciding on a vision of land bound to a specific era

---

[35]   Jonathan Z. Smith, "To Take Place," in Graham Harvey (ed.), *Critical Categories in the Study of Religion—Ritual and Religious Belief: A Reader* (London: Equinox Publishing, 2005), p. 33; my italics.

[36]   This is in contrast to the popular conception held by some that ritual is boring and a time to stop paying attention because "we've repeatedly done it before."

[37]   For a clear statement of these agricultural goals, see Wes Jackson (*Altars of Unhewn Stone: Science and the Earth* (San Francisco: North Point Press, 1987)). For an exploration of some of the ethics behind such agriculture, see Paul Thompson (*The Spirit of the Soil: Agriculture and Environmental Ethics* (New York: Routledge, 1995)).

[38]   Wes Jackson, *New Roots for Agriculture: New Edition* (Lincoln: University of Nebraska Press, 1985), pp. 6–7.

and then working to make the vision come to life. Or, as Edward Eisenberg puts it, "Manmade landscapes survive only at the sufferance of the wildness around them, or the wildness that remains in them."[39] Such sufferance brings with it a sense of shame that can be found in the humanly managed agricultural landscapes where this sufferance is daily engaged and acted out. In other words, we actively choose what our manmade landscapes are going to resemble. This choice entails killing; a historic and cultural vision of what the land will look like and be used for; and requires us to encounter limits and dependency upon the gifts of the Other (which we economically repay by visiting sufferance upon the wildness around us).

Given these considerations, an ideal context for an eco-agricultural restoration ritual can be a Community Supported Agriculture (CSA) farm, especially one that is relatively close to an urban area. This type of farming joins farmers with "shareholders" who pay a certain fee to a farmer at the beginning of a growing season. In essence, they enter into a contract with the farmer, who repays the shareholder with produce as the season progresses. Three benefits result from this arrangement: (1) ideally, the CSA supports sustainable agriculture ideals; (2) there is a built-in audience and membership for undertaking eco-rituals, as these can be performed on the property of the farmer and can be open to the participation of CSA members; and (3) this community is of a small enough size so that manageable rituals can be created and performed such that all members have a chance to observe and/or participate in the ritual/s.

Moreover, contemporary practitioners of sustainable agriculture are typically concerned with many current environmental issues: seed saving; creating and restoring habitat for numerous species, including native species; building soil fertility (in effect repaying an economic debt to nature); providing local food and contributing to local food systems; working with nature's systems; working with the seasons and being dependent on larger-than-human weather patterns; birthing, growing and killing flora and fauna; and managing landscapes for human use. These farmers must also deal with invasive species (including encroaching houses and urban sprawl); "pests"—whether insect or animal; creating compost; and deciding how to use nature to meet human caloric and aesthetic needs. Any and all of these aspects of sustainable agriculture, which must be engaged throughout the year, season to season, thus become occasions to perform ritual with members of the CSA and the surrounding community. Consistent with the argument of this chapter, these rituals will develop and engage USPs designed to promote sustainable land use. They will also explicitly recognize our dependency on the non-human world and the shame that results from this economic transaction.

A ritual could be built around digging a hole for a pond; birthing and/or slaughtering of animals; preparing of beds for the first planting of the season; harvesting of crops; building a greenhouse; transplanting plants into the field; turning compost; buying a new tractor; picking up of shares by CSA members;

---

[39] Evan Eisenberg, "Eden's Ecology," in Bron Taylor (ed.), *The Encyclopedia of Religion and Nature* (New York: Continuum, 2005), p. 573.

trapping and releasing a field mouse or raccoon or killing a deer; cutting and drying medicinal herbs, etc. Almost any and all aspects of a small-scale sustainable farm entail ritual, or an event that can be ritualized. The challenge lies in using these events within the economy of a farm's life as occasions to engage with shame via building and enacting a liturgically coded ritual.

Of equal import is that all of these activities require human intervention into processes of nature. In farming, we are manipulating, shaping and managing nature, attempting to turn it into an idealized landscape that can offer us calories. Also, rather than leaving this work to the less than 2 percent of the US population that engages in this process—and not to mention the many immigrant workers who process our food items in sub-standard working conditions—eco-rituals on a farm re-integrate the humans who are economically and calorically dependent on a foodshed into this process of management.

What matters in eco-rituals is both the context and the relationship between the performers and participants, for "it is within these relationships that the fundamental epistemological and ontological relations of any society are likely to be implicated and worked out: because this is the creative edge where reality is socially constructed."[40] What I am arguing, and what an eco-ritual undertaken at a sustainable farm could exemplify, is that "nature" is brought into our awareness as a participant in ritual. We are actively negotiating and constructing our social reality, our environmental ethics, and our relations with nature, and this construct is purposely intensified and delved into during the performance of eco-ritual. Both the performers and participants of an eco-ritual will deliberate upon the types of morality they wish to encode in the ritual; the reasons behind this (ideally the Logos of ecology); and will begin the hard work of constructing and then enacting a seasonal ritual that will performatively encode a localized environmental activity and ethic capable of addressing shame as a loci for value creation.

Schieffelin continues, writing that, "Performance is also concerned with … the creation of presence [and] through these presences, they alter moods, social relations, bodily disposition and states of mind."[41] Eco-ritual deals with the presence of shame, strives for short-lived *communitas* (the altering of social relations), and allows for the Logos of ecological sustainability to be enumerated via USPs that are sanctioned by participating in the ritual. These rituals also engage our bodies and emotions, the body and bodies of nature, and they shape and construct our morality and values as these relate to our local habitats. Although few would question the assertion that rituals play a key role in negotiating relationships among people, our dominant, contemporary Western expectations about rituals are that they do and will not concern themselves with human–nature relations. This cannot be further from the truth, as the experience of many cultures makes it

---

[40] Edward Schieffelin, "Problematizing Performance," in Graham Harvey (ed.), *Critical Categories in the Study of Religion—Ritual and Religious Belief: A Reader* (London: Equinox Publishing, 2005), p. 136.

[41] *Ibid.*, p. 125.

clear that this is true for relations between humans and their environment as well. To ignore this or to fail to take advantage of this is a serious mistake, and it is to the environmental movement's detriment that we do not utilize this performative technology to work towards a more effective environmental ethic and practice.

Thus, an email invitation can be sent to CSA members and other people in the larger community who want to be present for a performed ritual. The farmer/s can work with a local religious leader, or interfaith leaders committed to eco-theological ideals, and decide what action is to be ritualized, when, and where. Participants can arrive and encounter a "sacred space"—this might be a roped off area in which to park and then a trail to walk through to get to the ritual site. The burning of candles, sage and incense to cleanse the area and participants may then occur, coupled with the recitation of poems or singing of hymns. In other words, a ritualized eco-liturgy begins in some fashion, with clearly defined roles for both the leaders and audience participants—indeed, one of the challenges is to create a performative community recognized as such. Ideally a mood of solemnity is generated (this does not mean the fostering of a guilt-laden piety) so that the breach—our needed intrusion into the landscape or causing of death and thus a shameful, unequal economic exchange that carries with it moral obligation—is recognized and addressed. The ritual then generates and enumerates USPs based on an ecological understanding of the farm ecosystem, and the role of humans in managing the ecosystem. These USPs will explicitly contain a reference to ethics, and to "right actions" that attempt to embody these ethics. This can be a ritualized slaughter, a ritualized planting of seeds, or a ritualized removal of an invasive species. Any act on a farm that involves direct human intervention into the farm landscape; any act that requires the taking of flora or fauna; any act that includes us in the problematic aspect of being members of a larger-than-human community—all of these present themselves as opportunities to move from shame to beauty and community. As the theologian James Gustafson writes, "Moral ambiguity and even an element of the tragic are almost always present when we have good reasons to take life; the outcomes are costly to beings that have value for other beings and are worthy of respect because they exist."[42] This "costly" economic exchange is denied by a sentimental longing for a prelapsarian eco-age. Rather than this wishful thinking, eco-ritual embraces the tragic and ambiguous aspects of our community relations and provides for an occasion for moving toward transcendent values and the encoding of ecological USPs.

One important thing to remember is that these rituals are seasonal, so a liturgical calendar of performance can be created. Thus, a community of practitioners develops over time. Because of the emotional, physiological and physical aspects of the eco-ritual, these USPs have a better chance of influencing the values and practices of the community of practitioners than does a sentimental ethic that does not simultaneously engage feelings, brains and bodies. If local religious

---

[42]   James Gustafson, *A Sense of the Divine: The Natural Environment From a Theocentric Perspective* (Cleveland: The Pilgrim Press, 1994), p. 65.

leaders lead the rituals, then an extra level of cultural authority is added to the proceedings.[43] Local ecologists, scientists and the farmers themselves can also help lead the ritual, assuring that an ethic built on both religion and ecology can be generated. It is important to remember that the ritual and "homily," if one is indeed delivered during the performance, will address our shame and our radical dependency, including the debt we cannot repay to the planet. This is not puritanical harping on guilt nor an overly romantic and sentimentalized "eco-sin." Rather, it is a process of moving a community through acknowledgment of shame—not guilt, but shame—into an engagement with higher values like beauty, appreciation and respect for the processes of nature we depend upon for our survival. This is because an eco-ritual magnifies and accentuates our needed hands-on (and shovel-on, hoe-on or tractor-on) transgressions into the economy and lifeforms of a farm ecosystem.

Conclusion

I have outlined Bill Jordan's theory of shame-to-value and The Values Project Study Group's argument for the role of performative ritual in generating a non-sentimental, effective environmental ethic rooted in place. In this turn to ritual it is important to keep in mind that all religion and religious rituals are social constructs.[44] This does not imply that religion, and religious rituals, are not functionally efficacious in peoples' lives. Rather, it means that any ritual, including an eco-ritual, is a social creation that has an origin at some place and time, with some person or group of people. It is not a strained leap of imagination to see how many of the current NGOs that organize environmental clean-ups and beach sweeps or the removal of invasive species, or the religious bodies that are generating eco-theological liturgies, can move into generating USPs based on the recognition of shame. With this move, we proceed past the sentimental environmentalism characteristic of the last century or so of our environmental thinking and enter into the possibility of generating values of community, beauty, and a constrained, localized, ecologically nuanced ethic.

This nuanced ethic has an unrecognized potential to be highly effective, especially if a liturgical community develops over time. This is because ritual is a technology that both creates community and fosters lasting ethical obligations to local places. My exploration of Roy Rappaport and both Victor and Frederick Turner suggests that place is central to the performance of ritual and that the performance of ritual attempts to deal with breaks in community cohesion (both human–human and human–nature communities).

---

[43]   Given that some polls report that up to 90 percent of all Americans believe in some form of deity, religious authority carries extra significance in the United States.

[44]   Russell McCutcheon, *Manufacturing Religion: The Discourse on Sui Generis Religion and the Politics of Nostalgia* (New York: Oxford University Press, 1997).

Based on personal experience and years of study, I am comfortable in arguing that ritual *must* enter into our environmentalisms as a necessary ingredient if we are to have healthy communities and healthy landbases. This holds even more as we enter into decades of extreme flux and migration due to climate destabilization. Such flux presents challenges to the goals of restoration ecology, sustainable agriculture and many other environmentalisms.

To close, I recognize that this call is a serious challenge to a highly anthropocentric, puritan, individualistic, consumer culture that abstracts our bodies, minds, emotions and financial market economy out of nature. Nonetheless, the USPs, the creation of *communitas*, and the engagement with shame that all entail to the performative technology of eco-restoration ritual/s are epistemological and ontological tools we need if we are to recreate, replace and restore a damaged world.

Bibliography

Abram, David, *The Spell of the Sensuous: Perception and Language in a More-Than-Human World* (New York: Vintage, 1997).

Baker, John R. and Michael Winkelman, eds, *Supernatural as Natural: A Biocultural Approach to Religion* (Upper Saddle River: Pearson Prentice Hall, 2010).

Bell, Catherine, "Performance," in Mark Taylor (ed.), *Critical Terms for Religious Studies* (Chicago: University of Chicago Press, 1998), pp. 205–24.

Bellah, Robert, "Meaning and Modernity: American and the World," in Richard Madsen, William M. Sullivan, Ann Swidler and Steven M. Tipton (eds), *Meaning and Modernity: Religion, Polity, and Self* (Berkeley: University of California Press, 2002), pp. 255–76.

Commoner, Barry, *The Closing Circle: Nature, Man, and Technology* (New York: Random House Publishing, 1971).

Crouch, David, "Performances and Constitutions of Natures: A Consideration of the Performance of Lay Geographies," in Bronislaw Szerszynski, Wallace Heim and Claire Waterton et al. (eds), *Nature Performed: Environment, Culture and Performance* (Oxford: Blackwell Publishing/The Sociological Review, 2003), pp. 17–30.

DeLoria, Jr., Vine, *God is Red: A Native View of Religion* (Golden, CO: Fulcrum Publishing, 1994).

Eisenberg, Evan, "Eden's Ecology," in Bron Taylor (ed.), *The Encyclopedia of Religion and Nature* (New York: Continuum, 2005), pp. 572–5.

Grimes, Ronald, *Ritual Criticism: Case Studies in Its Practice, Essays in Its Theory* (Columbia: University of South Carolina Press, 1990).

Grimes, Ronald, "Ritual," in Willi Braun and Russell McCutcheon (eds), *Guide to the Study of Religion* (New York: Cassell, 2000), pp. 259–70.

Grimes, Ronald, "Ritual," in Bron Taylor (ed.), *The Encyclopedia of Religion and Nature* (New York: Continuum, 2005), pp. 1385–8.

Gustafson, James, *A Sense of the Divine: The Natural Environment From a Theocentric Perspective* (Cleveland, OH: The Pilgrim Press, 1994).

Jackson, Wes, *New Roots for Agriculture: New Edition* (Lincoln, NB: University of Nebraska Press, 1985).

Jackson, Wes, *Altars of Unhewn Stone: Science and the Earth* (San Francisco: North Point Press, 1987).

Jordan III, William, *The Sunflower Forest: Ecological Restoration and the New Communion with Nature* (Berkeley: University of California Press, 2003).

Loy, David R., "The Religion of the Market," *Journal of the American Academy of Religion* 65 (1997): pp. 275–90.

McCloskey, David, "What Community is Not: Three Myths about Community" (unpublished manuscript, Seattle University, 1997).

McCutcheon, Russell, *Manufacturing Religion: The Discourse on Sui Generis Religion and the Politics of Nostalgia* (New York: Oxford University Press, 1997).

Neusner, Jacob, ed., *The Christian and Judaic Invention of History* (Atlanta: Scholars Press, 1990).

Peterson, Anna L., *Being Human: Ethics, Environment, and Our Place in the World* (Berkeley: University of California Press, 2001).

Quinn, Daniel, *Ishmael: An Adventure of the Mind and Spirit* (New York: Bantam/ Turner, 1992).

Quinn, Daniel, *The Story of B: An Adventure of the Mind and Spirit* (New York: Bantam/Turner, 1996).

Rappaport, Roy, *Ecology, Meaning, and Religion* (Berkeley: North Atlantic Books, 1979).

Rappaport, Roy, *Ritual and Religion in the Making of Humanity* (New York: Cambridge University Press, 1999).

Schieffelin, Edward, "Problematizing Performance," in Graham Harvey (ed.), *Critical Categories in the Study of Religion—Ritual and Religious Belief: A Reader* (London: Equinox Publishing, 2005), pp. 124–38.

Smith, Jonathan Z., "To Take Place," in Graham Harvey (ed.), *Critical Categories in the Study of Religion—Ritual and Religious Belief: A Reader* (London: Equinox Publishing, 2005), pp. 26–50.

Bronislaw Szerszynski, Wallace Heim and Claire Waterton, eds, *Nature Performed: Environment, Culture and Performance* (Oxford: Blackwell Publishing/The Sociological Review, 2003).

Thompson, Paul, *The Spirit of the Soil: Agriculture and Environmental Ethics* (New York: Routledge, 1995).

Tuan, Yi-Fu, "Discrepancies Between Environmental Attitude and Behaviour: Examples from Europe and China," in David Spring and Eileen Spring (eds), *Ecology and Religion in History* (New York: Harper and Row, 1974).

Turner, Frederick, *Beyond Geography: The Western Spirit Against the Wilderness* (New York: Viking Press, 1980).

Turner, Frederick, *Rebirth of Value: Meditations on Beauty, Ecology, Religion, and Education* (New York: SUNY Press, 1991).

Turner, Victor, "Are There Universals of Performance in Myth, Ritual, and Drama?" in Richard Schechner and William Appel (eds), *By Means of Performance: Intercultural Studies of Theatre and Ritual* (New York: Cambridge University Press, 1990), pp. 8–18.

Chapter 4

Recreating [in] Eden:
Ethical Issues in Restoration in Wilderness

Daniel T. Spencer

Introduction

Picture each of the following scenarios: a helicopter hovers over a river in Virginia's Saint Mary's Wilderness, releasing a load of lime to buffer acidity in the stream caused by acid rain; a Forest Service crew applies acres of 2,4D herbicide to hillsides covered with the invasive weed, spotted knapweed, in the Frank Church River of No Return Wilderness in Idaho; a single engine aircraft drops low over a lake in Montana's Bob Marshall Wilderness Area and releases hundreds of gallons of rotenone, a piscicide designed to kill nonnative fish in lakes that were stocked early in the twentieth century; a water truck rumbles across the desert of the Kofa Wilderness in Arizona to refill underground reservoirs designed to provide water to endangered desert bighorn sheep; government nurseries raise thousands of rust-resistant whitebark pine seedlings to reforest alpine areas in the Lee Metcalf Wilderness ravaged by whitebark pine blister rust; Forest Service crews set controlled burns in dense forest in the Scapegoat Wilderness, overgrown from a century of fire suppression.[1]

Each of these examples, and many more beyond them, have either occurred or have been proposed in congressionally designated wilderness areas in an effort to restore or maintain perceived natural conditions, now being degraded by modern human actions, both direct and indirect. Wilderness is both a modern designation and a primeval ecological reality; as a symbol of places largely untouched by modern human influence, wilderness for many is perhaps closest to the biblical image of Eden as a paradisiacal place uncreated by humans and unharmed by us. "Re-creating" and "Recreating in" Eden serve as biblical metaphors for returning to a predisturbance harmonious state—in many ways what proposals for restoration in wilderness seek.

Increasingly the science of restoration ecology and the techniques of ecological restoration provide wilderness managers with tools that can be used to restore ecological integrity in wilderness, but neither the science nor the techniques can

---

[1] P. B. Landres, M. W. Brunson and L. Merigliano, "Naturalness and Wildness: The Dilemma and Irony of Ecological Restoration in Wilderness," *Wild Earth* 10 (2001): pp. 77–82.

answer the host of philosophical and ethical issues that arise as to whether we *should* undertake restoration in wilderness. In this chapter, I do not address that question directly, but rather use it to introduce a model for ethical decision-making about restoration in wilderness developed recently by students in my graduate course at the University of Montana, "Ethical Issues in Ecological Restoration." The model is designed to give wilderness managers and other interested persons a process by which to consider not only scientific and legal issues at stake in restoration in wilderness, but also moral and ethical issues and perspectives that arise. After explaining the model, I walk through two case studies—removing nonnative fish and introducing native fish in the Bob Marshall Wilderness in Montana, and providing water guzzlers to maintain declining desert bighorn sheep populations in the Kofa Wilderness in Arizona—to illustrate how the model might work. Through this model we hope to introduce a participatory deliberative process that can involve a wide net of stakeholders and reduce conflicts and litigation over decisions and actions by wilderness managers involving restoration in wilderness.

A Central Question: What Counts as *Good* Ecological Restoration?

Exploring the ethical dimensions of restoration quickly leads one to thinking about what counts as *good* ecological restoration, in *what* contexts, and *why*. The person who perhaps has explored this issue in most depth and rigor is the Canadian philosopher Eric Higgs, himself a past president of the Society for Ecological Restoration. In his seminal 1997 article, "What is Good Ecological Restoration?" Higgs nests *effective* restoration within expanding ellipses that integrate both economic and social-historical factors that must be considered for restoration also to count as *good* restoration.[2]

Higgs develops this model further in his 2003 book, *Nature By Design: People, Natural Process, and Ecological Restoration*. He posits four keystone concepts of good ecological restoration: ecological integrity, historical fidelity, focal restoration, and wild design. Ecological integrity, for Higgs, is "an all-encompassing term for the various features … that allow an ecosystem to adjust to environmental change."[3] It incorporates the idea of recovering previous natural conditions, which leads to the second keystone concept, historical fidelity: "loyalty to predisturbance conditions, which may or may not involve exact reproduction."[4] Together ecological integrity and historical fidelity encompass the inner circle of effective restoration and ecological fidelity in Higgs' 1997 model.

---

[2]   E. S. Higgs, "What is Good Ecological Restoration?" *Conservation Biology* 11(1997): pp. 338–48.

[3]   E. S. Higgs, *Nature by Design: People, Natural Process, and Ecological Restoration* (Cambridge, MA: MIT Press 2003), p. 122.

[4]   *Ibid.*, 127.

Higgs is concerned, however, with the contemporary context of increasing technological efficiency and commodification that threatens to sever restoration from its roots as a community practice. He draws on the work of philosopher Albert Borgmann and his emphasis on "focal practices"—activities that focus persons and communities on things that are deeply engaging and can ground one in systems of meaning rooted in history and place—to argue that "focal restoration" is also critical to good restoration.[5] Finally he explores the paradox that restoration requires intentional manipulation of natural or wild processes, and posits "wild design" to designate the design process of trying to return or increase the independent character of wildness to ecosystems and landscapes.

The Wilderness Management Dilemma: Framing the Issue

Because of the legally defined nature of congressionally designated wilderness in the 1964 Wilderness Act, ethical issues of wilderness and restoration primarily focus on the first two of Higgs' keystone concepts, ecological integrity and historical fidelity.[6] We can frame the discussion in terms of what Landres et al. called the "central wilderness management dilemma":[7] "Increasingly, wilderness managers must choose between the objective of wildness ('untrammeled' wilderness) and the objectives of naturalness and solitude."[8] Hence when considering whether or not to engage in ecological restoration in wilderness, there are two primary bounding conditions: (1) our increasing *proficiency* at restoration:

---

[5]   A. Borgmann, *Technology and the Character of Contemporary Life* (Chicago: University of Chicago Press, 1984); W. Throop and R. Purdom, "Wilderness Restoration: The Paradox of Public Participation," *Restoration Ecology* 14 (2006): pp. 493–9.

[6]   For a discussion of tensions around focal restoration in wilderness, see Throop and Purdom, "Wilderness Restoration" and E. S. Higgs, "Restoration Goes Wild: A Reply to Throop and Purdom," *Restoration Ecology* 14 (2006): pp. 500–3. For further development of the concept of wild design, see E. S. Higgs and R. Hobbs, "Wild Design: Intention, intervention, reciprocity" in D. Cole and L. Yung (eds), *Conserving Parks and Wilderness in the 21st Century: Naturalness and Beyond* (Washington, DC: Island Press, 2009).

[7]   P. B. Landres, P. W. White, G. Aplet and A. Zimmermann, "Naturalness and Natural Variability: Definitions, Concepts, and Strategies for Wilderness Management," in D. L. Kulhavy and M. H. Legg (eds), *Wilderness and Natural Areas in Eastern North America: Research, Management, and Planning* (Nacoghoches, TX: Center for Applied Studies in Forestry, Stephen F. Austin State University, 1998a), pp. 41–50.

[8]   D. N. Cole and W. E. Hammit, "Wilderness Management Dilemmas: Fertile Ground for Wilderness Management Research," in D. N. Cole, S. F. McCool, W. A. Freimund and J. O'Loughlin (eds), *Wilderness Science in a Time of Change Conference—Volume 1: Changing Perspectives and Future Directions; 1999 May 23–27; Missoula, MT. Proceedings RMRS-P-15-VOL-1* (Ogden, UT: U.S. Department of Agriculture, Forest Service, Rocky Mountain Research Station, 2000), p. 58; D. N. Cole, "Management Dilemmas that will Shape Wilderness in the 21st Century," *Journal of Forestry* 99 (2001): pp. 4–8.

often we *can* restore ecological integrity to degraded wilderness areas, but *should* we? And (2) the 1964 Wilderness Act that mandates managing wilderness for *both* naturalness *and* wildness or untrammeledness;[9] actual management decisions about restoration, however, often mean that one of these qualities is compromised in favor of upholding the other.

Here is how Peter Landres of the Aldo Leopold Wilderness Research Institute frames these issues.[10] Currently the historically *natural* conditions of many wilderness areas are being degraded by modern human influences. Examples include the spread of nonnative invasive species that threaten the complexity and biodiversity of native ecosystems, the accumulation of unnatural fuel loads in forests from a century of fire exclusion, the fragmentation of habitat leading to the local extirpation and extinction of species, and many more. Hence the question "We *could* try to restore natural conditions, but *should* we?" The tension inherent in this question is embedded in the very definition of wilderness in the Wilderness Act of 1964.[11] In Section 2(c), wilderness is defined legally: "A wilderness … is hereby recognized as an area where the earth and its community of life are untrammeled by man" (Wilderness Act, Sec 2(c)). Howard Zahniser, primary author of the Wilderness Act, was particularly concerned about the human temptation to "improve" wilderness through management, and he chose the word "untrammeled" carefully, arguing that the inspiration for wilderness preservation "is to use 'skill, judgment, and ecologic sensitivity' for the protection of some areas within which natural forces may operate without man's management and manipulation."[12] Yet the same section continues: "An area of wilderness is further defined to mean … an area … retaining its primeval character … which is protected and managed so as to preserve its natural conditions" (Wilderness Act, Sec 2(c)). "Natural" here has been interpreted in subsequent documents to mean the species, patterns and processes that evolved in the area.[13]

---

[9]   P. B. Landres, R. L. Knight, S. T. A. Pickett and M. L. Candenaso, "Ecological Impacts of Administrative Boundaries," in *Stewardship Across Boundaries* (Washington, DC: Island Press, 1998), pp. 39–64.

[10]   Peter Landres, personal communication, October 2008.

[11]   For the purposes of this chapter I sidestep the many important philosophical debates about the concept and definition of wilderness itself, and focus instead on the statutory definition of the Wilderness Act. For philosophical and conceptual debates about wilderness, see W. Cronon, *Uncommon Ground: Rethinking the Human Place in Nature* (New York: W. W. Norton, 1996) and J. B. Callicott and M. P. Nelson, *The Great New Wilderness Debate* (Athens, GA: University of Georgia Press, 1998).

[12]   Zahniser 1963; cited in P. B. Landres, S. Boutcher, L. Merigliano, C. Barns, D. Davis, T. Hall, S. Henry, B. Hunter, P. Janiga, M. Laker, A. McPherson, D. S. Powell, M. Rowan, and S. Sater, "Monitoring Selected Conditions Related to Wilderness Character: A National Framework," Gen. Tech. Rep. RMRS-GTR-151 (Fort Collins, CO: U.S. Department of Agriculture, Forest Service, Rocky Mountain Research Station, 2005).

[13]   Landres et al., "Monitoring Selected Conditions."

Hence the Wilderness Act requires that we manage wilderness to protect two core wilderness values, naturalness and untrammeledness. In our current situation where much of wilderness is being degraded, taking no restoration action preserves the untrammeled character of wilderness, but leaves the natural character degraded and low. Conversely, taking restoration actions can restore ecological integrity and naturalness, but often only by "trammeling" wilderness in the process, intentionally manipulating processes and elements of and in wilderness areas. This, then, is the wilderness management dilemma: taking restoration action compromises the *untrammeled* value of wilderness, but *not* taking action may compromise the *natural* value of wilderness. In many cases this tension is a true ethical dilemma: maintaining both naturalness and untrammeledness are central ethical values, yet at least one and maybe both will be compromised, if not violated in reaching a decision.

Developing the Ethical Framework

This is the background the class used to develop a framework for working through ethical dilemmas around restoration. In recent years many managerial decisions about restoration in wilderness have triggered tremendous acrimony in the wilderness and conservation communities, often ending up being litigated in court. Our goal, therefore, was to develop a model that might help managers and other wilderness stakeholders work through a particular issue, and have ethical considerations inform the decision-making—hence, a practical, guided process. The audience therefore includes wilderness managers and other stakeholders interested in restoration in wilderness.

To begin, we decided to use a mixed contextual approach that combines attention to both ethical principles and consequences; hence it is neither a strictly deontological nor consequentialist approach. The framework is based loosely on two models: (1) a Spiral of Praxis model of ethics that works through five main steps:

1. Setting the Context and Naming the Relevant Community for Ethical Analysis
2. Understanding the Historical-Social Context and Roots of the Issues
3. Engaging in Deliberate Ethical Analysis and Reflection
4. Planning for Engagement in Light of Ethical Values and Consequences
5. Choosing the Best Option and Justifying it.

(2) A second model for decision-making comes from bioethics, developed by ethicist Ron Hamel:

1. Gather Information
2. Identify the Issue

3. Review Core Commitments
4. Identify Alternatives
5. Assess Alternatives in Light of Core Commitments and Relevant Values
6. Make a Decision and Justify Your Choice
7. Ongoing Evaluation of Decision and Actions.

Here is the outline of the model. In the following section, I highlight some of the important features for how they relate to restoration in wilderness.[14]

1. Gather Information
   a. What aspect of the wilderness is in jeopardy and how does the current situation compromise wilderness character? Agency seeks out and reviews scientific literature, current public involvement, precedents.
   b. What wilderness values are at risk?
   c. Who has a stake in it?
2. Decide on Facilitation
   a. Will the manager role be compatible with the facilitator role?
   b. Identify potential facilitators (could be the wilderness manager, another agency employee or someone outside of the agency).
3. Form a Working Group
   a. Invite known stakeholder groups to contribute representatives.
   b. Review definitions of ethics and restoration.
   c. Articulate a memorandum of understanding (MOU).
   d. Decide whether an outside facilitator is needed.
4. Review core ethical commitments. Parties involved establish familiarity with:
   a. Wilderness Act—to ensure ecosystem integrity.
   b. National Environmental Policy Act (NEPA)—to ensure decision-making integrity.
   c. The Precautionary Principle—to ensure a practical minimum standard for acceptable actions.

---

[14] Since this article was originally written, a special issue of *Environmental Management* has been published, focusing on "Conservation without Borders: Building Communication and Action across Disciplinary Boundaries for Effective Conservation." Of particular relevance to issues in this chapter are papers by S. W. Margles, R. B. Peterson, J. Ervin and B. A. Kaplin, "Conservation Without Borders: Building Communication and Action Across Disciplinary Boundaries for Effective Conservation," *Environmental Management* 45 (2010): pp. 1–4; J. S. Gruber, "Key Principles of Community-Based Natural Resource Management: A Synthesis and Interpretation of Identified Effective Approaches for Managing the Commons," *Environmental Management* 45 (2010): pp. 52–66; and T. A. Muñoz-Erickson, B. Aguilar-González, M. R. R. Loeser and T. D. Sisk, "A Framework to Evaluate Ecological and Social Outcomes of Collaborative Management: Lessons from Implementation with a Northern Arizona Collaborative Group," *Environmental Management* 45 (2010): pp. 132–44.

d. Tribal sovereignty and traditional ecological knowledge—to ensure political integrity and historical fidelity.
e. Case specificity—to ensure viability, significance and respect within the local context.

5. Review Ethical Issues of the Case
   a. What is the economic, social, ecological, cultural and historical significance of the area?
   b. What human values or perceptions of the area exist?
   c. What is the dilemma?
   d. What are the positions of working group members, their priorities and their ethics?
   e. Are there conflicts between ethical positions?
   f. Are there conflicts, which involve agency core ethical commitments?

6. Generate Alternatives
   a. Generate and describe alternatives, including the no-action alternative.
   b. What are the likely positive and negative outcomes of these options in the long and short term?

7. Examine Options and Choose Best Course of Action
   a. Do any methods or outcomes of the alternatives conflict with the core ethical commitments?
   b. Revisit the dilemma: what philosophical considerations underlie it?
   c. Does any proposed action wholly address the situation? *If not* ...
   d. Which compromise would best resolve the conflict? Why?

Significant Features of the Model

The model is premised on five core ethical commitments that are reviewed in step 4. Each commitment contains core values that should ground decision-making around restoration in wilderness and be upheld to the greatest extent possible in any particular case. Some principles stem from legal requirements and statutes and I discuss here how they reflect important ethical considerations as well.[15]

- The first ethical commitment is to the integrity of wilderness, and is informed by the Wilderness Act and the particular enabling legislation of the wilderness area where action is proposed. In its characterization of wilderness as natural, untrammeled, undeveloped and able to provide solitude, the Wilderness Act and related legislation create a goal of preserving the wilderness character of an area for any action.[16]

---

[15] The wording in this section is from the class paper and reflects the primary authorship of Oskar Coles.

[16] P. Landres, C. Barns, J. Dennis, T. Devine, P. Geissler, C. McCasland, L. Merigliano, J. Seastrand, and R. Swain, "Keeping it Wild: An Interagency Strategy to Monitor Trends

- The second ethical commitment is to the participatory process, and is informed by the National Environmental Policy Act (NEPA).[17] NEPA's insistence on transparency and democratic procedure shapes the decision-making process. Due to the specificity that arises in each case, it is also important to allow for public participation in making informed, defensible ethical decisions.
- The third ethical commitment is to foresight, and is informed by the Precautionary Principle, which states, "In the absence of a scientific consensus that harm will not ensue, the burden of proof lies falls on those who would advocate taking the action."[18] Since neither the Wilderness Act nor NEPA makes explicit any standard for acceptable action, the Precautionary Principle provides a practical minimum standard for the decision.
- The fourth ethical commitment is to tribal sovereignty and traditional ecological knowledge (TEK), as foundations for understanding historical fidelity and traditional land use.[19] Thus, respect for humanity, acknowledgment of history, and dedication to inclusiveness and justice in future actions are brought to the spirit of the decision.
- The fifth ethical commitment is to case-specificity, and is informed by current social context, local history of land use, ecological processes and overlapping agency management. This commitment acknowledges the necessity to involve local communities, consider local circumstances and collaborate with all responsible agencies. Thus, a decision will be locally viable, meaningful and respectful of the context in which it is made.

A word about process. The framework outlined here is intended to guide a highly intensive collaborative process involving multiple stakeholders. Ideally a group of stakeholders would form not for just one specific case, but as a body that knows and cares deeply about a particular wilderness area, develops a history with the management of that area, and can bring a working knowledge of the wilderness and the group to examining specific cases. Hence care should be taken to form a working group that is inclusive and the process transparent and participatory. In some cases it may be helpful to have an outside facilitator to guide the process,

in Wilderness Character across the National Wilderness Preservation System," Gen. Tech. Rep. RMRS-GTR-212. (Fort Collins, CO: U.S. Department of Agriculture, Forest Service, Rocky Mountain Research Station, 2008).

[17]   For information on the National Environmental Policy Act (NEPA), see www.epa. gov/compliance/nepa/index.html and http://ceq.hss.doe.gov/nepa/nepanet.htm

[18]   C. Raffensperger and J. Tickner, *Protecting Public Health and the Environment: Implementing the Precautionary Principle* (Washington, DC: Island Press, 1999).

[19]   A. Watson, L. Alessa, and B. Glaspell, "The Relationship between Traditional Ecological Knowledge, Evolving Cultures, and Wilderness Protection in the Circumpolar North," *Conservation Ecology* 8 (2003): p. 2.

particularly when managers and agency personnel may be perceived as interested stakeholders rather than neutral facilitators. Steps 5–7 are particularly important for exploring the range of ethical issues and generating and examining many different alternatives. Assessing the alternatives in light of (1) likely consequences, both short- and long-term, and (2) core commitments and values, increases the likelihood of coming up with a course of action that minimizes conflict and best reflects the overall values of the working group and the general public.

Two Case Studies

In the remainder of the chapter, I outline two case studies that illustrate some of the complexities of the wilderness management dilemma, and how the ethical framework can be used to work through the issues. Given the scope of this chapter, the analysis is necessarily abbreviated, but I hope to illustrate some of the main issues.[20]

Restoring Native Fish in the Bob Marshall Wilderness Area, Montana

The current proposal for restoration of westslope cutthroat trout (WCT)—currently listed as a "species of concern"—to lakes and streams in the South Fork of the Flathead River watershed has become very controversial as it includes poisoning lakes with nonnative fish and restocking these historically fishless lakes with WCT in the Bob Marshall Wilderness. Is it ethically justifiable to compromise both the central wilderness values of "naturalness" and "untrammeledness" for the meritorious end of expanding pure WCT habitat in order to increase chances of survival of this species—even if it includes wilderness habitat with no historical record of hosting these native fish?

In May of 2006, the Bonneville Power Administration (BPA) signed a Record of Decision (ROD) to do just this, announcing the BPA's intent to fund the South Fork Flathead Watershed Westslope Cutthroat Trout Conservation Program.[21] The South Fork drainage is home to one of the largest known genetically pure populations of WCT,[22] but this population is threatened by hybridization with

---

[20]   In the following sections, the text in quotation marks is drawn from the graduate student paper that was the culmination for the course project on developing the framework for ethical analysis. Contributors to this paper are listed in the Acknowledgments at the end of the chapter. An edited version of this much longer paper is being prepared for publication with the United States Forest Service.

[21]   Bonneville Power Administration. 2006. South Fork Flathead Watershed Westslope Cutthroat Trout Conservation Program, Record of Decision. Retrieved February 7, 2009 from http://regulations.vlex.com/vid/flathead-westslope-cutthroat-trout-22139807: p. 1.

[22]   Bonneville Power Administration. 2005. South Fork Flathead Watershed Westslope Cutthroat Trout Conservation Program, Final Environmental Impact Statement (DOE/EIS-

nonnative trout that were introduced to the watershed from the 1920s through the 1940s in order to increase angling and recreation opportunities.[23] This included several historically fishless lakes in what later was designated the Bob Marshall Wilderness Area (BMWA)—and these lakes serve as headwaters for the South Fork of the Flathead. The BPA's conservation plan involves two distinct steps: (1) using a piscicide to poison the lakes to remove the nonnative fish populations, and then (2) restocking the lakes with hatchery-raised genetically pure WCT. The hope is that this introduced native fish population will expand traditional WCT habitat and prevent hybridization with nonnative fish. Implementing the plan will require using motorized vehicles and chemical fish toxicants within a congressionally designated wilderness area. Hence conflicting values reflect the complexity of this wilderness ethical dilemma: the core values of wilderness management internal to the BMWA—particularly preserving the area's natural and untrammeled qualities—conflict directly with actions needed to preserve a valued native species and its habitat outside the BMWA.

Briefly reviewing the core commitments of the model reveals some of the ethical complexity. In terms of the *commitment to wilderness integrity*, the proposed fish restoration project in the BMWA would potentially violate four core values of the Wilderness Act: managing wilderness so as to preserve its natural, untrammeled and undeveloped character, while maintaining opportunities for solitude. The *commitment to participatory process* has been disputed by several wilderness groups, resulting in litigation to try to reverse the proposed project. The Forest Service followed the *NEPA process* as part of its regulatory obligation, but many affected stakeholders do not feel their input was heard or incorporated, suggesting this commitment may have been met formally, but without honoring the intended spirit.[24] *Commitment to foresight as informed by the Precautionary Principle* reveals further complexity: do we honor the precautionary principle by proactively expanding native fish habitat into historically fishless waters in wilderness in hopes of increasing the chances of survival of the WCT? Or do we violate the precautionary principle by proposing an action in wilderness whose impacts are uncertain, using means that violate each of the core values of wilderness, if only in the short term? *In terms of commitment to tribal sovereignty and TEK*, this area borders on the Flathead Reservation where the Salish, Kootenai and Pend d'Oreille tribes have been active in restoration efforts. Tribal members generally support the efforts to restore the native WCT, though they differ on what

0353). Retrieved February 7, 2009 from http://www.gc.energy.gov/NEPA/draft-eis0353. htm: pp. 1–2.

[23]   Bonneville Power Administration. 2004. South Fork Flathead Watershed Westslope Cutthroat Trout Conservation Program, Draft Environmental Impact Statement (DOE/EIS-0353). Retrieved February 7, 2009 from http://www.gc.energy.gov/NEPA/draft-eis0353. htm: p. 7.

[24]   See "Cutthroat Business: Project to Poison Fish in Western Montana Lakes Advances," *The Spokane Spokesman-Review*, July 16, 2006.

are the appropriate means. In terms of *case and location specific commitments*, Montana state law mandates that the Montana Fish, Wildlife and Parks manage wildlife, fish and game in a manner that prevents the need for listing species under either the state or federal Endangered Species Acts.

The graduate students working on this case identified not one, but two primary ethical dilemmas:

1. First, unless nonnative trout populations in the BMWA are eliminated, the genetically pure and threatened WCT population in the South Fork watershed remains threatened with hybridization and competition. Yet eliminating the nonnative fish population in the BMWA inherently compromises the *untrammeled* state tenet of the Wilderness Act: at its core it involves several modes of human intervention into natural processes in wilderness.
2. Second, once the nonnative fish are removed, restocking the lakes in wilderness would serve to expand the viable native WCT population in the South Fork watershed, but since these lakes were historically fishless, restocking them compromises the *natural condition* mandate of the Wilderness Act by introducing a novel element into the native ecosystem, as well as a further *trammeling* of wilderness to carry out the restocking.

Hence the wilderness management dilemma in this case consists of three related but separate components that can be evaluated in terms of their ethical values: (1) *removing* nonnative fish from lakes and streams (and at least initially restoring their natural condition); (2) the *means* of fish removal—how best to accomplish this; (3) the decision to *restock* the lakes with genetically pure WCT. The goal of the removal process is to eliminate the threat of nonnative hybridization and competition with downstream WCT populations. The goal of the restocking process is two-fold: to provide a source of genetically pure WCT to seed downstream areas *and* to continue to provide traditional angling opportunities that predate designation of this area as wilderness. The conflicts inherent in these actions raise the question of the *value* of maintaining a genetically pure population of WCT, and whether there are other options that do not involve the trammeling of wilderness? If restoring what is *natural* in the BMWA is one of the desired ends and values, then restocking the lakes should not be part of the means, since it compromises *both* the *natural state* and *untrammeled condition* of wilderness.

Moving to identifying and assessing alternatives, the students noted that "the dilemma within the dilemma is that no single compromise can fully resolve" the ethical conflicts in this case. "Taking action goes against at least some values of wilderness. Not taking action goes against the need to protect a viable species of concern, promoting ecological integrity, and allows a non-natural condition in wilderness to persist—that is, lakes with introduced nonnative fish." When one combines the question of *means* to the two options of *removing* nonnative fish and *restocking* with WCT, there are many different potential options. Here I briefly review five main options.

Option 1: *use piscicide to remove nonnative fish from the wilderness lakes and do not restock with WCT*. This option would employ aircraft and motorized rafts to apply the fish toxins rotenone and antimycin to 11 wilderness lakes until the nonnative fish populations were eliminated. The lakes would then be left in their historically fishless condition and would not be restocked. *Pros* of this option include protecting the native WCT population from threatened hybridization and restoring the lakes to their historically fishless condition and hence a more natural ecosystem. *Cons* include short-term trammeling of wilderness and reducing recreational opportunities for anglers and outfitters. In terms of *ethical considerations*, since restoring ecological integrity is central to the ethical framework, one could argue that taking some action, even in wilderness, to ensure the survival of a threatened species would be supported—particularly if it is a short-term trammeling with few long-term effects. Hence this alternative compromises the "untrammeled" value of wilderness in order to restore its "natural" condition.

Option 2: *use piscicide to remove nonnative species from lakes and restock lakes with genetically pure WCT*. Fish removal methods would be the same as in Option 1, and then the lakes would be restocked with native fish. *Pros*: in addition to the benefits of removing nonnative fish, restocking the lakes would provide a source of WCT to seed downstream populations and would continue traditional angling and outfitting opportunities at wilderness lakes. *Cons* include those outlined in Option 1 plus the further compromising of the natural character of wilderness by introducing a fish species not historically present. In terms of *ethical considerations*, this option compromises *both* the "natural" and "untrammeled" values of wilderness for the sake of strengthening the threatened WCT population in the South Fork drainage. This then raises the question of the true value of restocking these lakes once the nonnative fish have been removed: is removing the nonnative fish enough action to remove the threat of WCT extinction, or is the restocking of the lakes absolutely necessary to prevent this?

Options 3 and 4: these address the question of *means*, and require the fish removal and restocking to be done without use of motorized aircraft and rafts, and without using chemical fish toxins. The *pros* of these options are removing unknown short- and long-term consequences of the piscicides in the lakes and surrounding ecosystems, and reducing the trammeling of wilderness through eliminating motorized access. The *cons* include a likely far greater cost and length of time to complete the fish removal, with the probability of less effectiveness and hence increased likelihood of a continued threat from the ongoing presence of nonnative fish. In terms of *ethical considerations*, the students noted that with these options "accepting that there are both known and unknown effects of piscicide use on other species in the ecosystem, looking further into the possibilities of alternative removal techniques, even if those techniques are more labor-intensive and costly, is ethically warranted to preserve the naturalness of wilderness."

Option 5: the final option is to Take No Action: "Due to necessary and unacceptable compromise of the values of wilderness with any action, no action should be taken in the BMW lakes, and nature will be left to take its course in those lakes. Lakes outside of the wilderness could still be treated." The *pros* of the Take No Action are that no further trammeling of wilderness occurs, and the risks of using piscicide in wilderness are avoided. The *cons* include leaving nonnative fish species in wilderness that will continue to pose a threat to the downstream native WCT population; a nonnatural condition in wilderness lakes persists. In terms of *ethical considerations*, this option upholds the untrammeled quality of wilderness, but at the expense of allowing a nonnatural condition to persist, and one that in addition threatens the well-being of a native species. The Precautionary Principle may best be upheld here, allowing wilderness to adapt on its own without further human manipulation.

After weighing each of these options in terms of their ethical pros and cons, the students came to consensus that Option 1 was the most acceptable option involving the least number of conflicts between the multiple management goals and ethical values at stake here. Option 1—removing the nonnative fish from the lakes but *not* restocking them with WCT—mitigates for past human mistakes and removes the threat of hybridization with downstream WCT. Not restocking the lakes with WCT avoids further trammeling of wilderness, and respects the value of naturalness in wilderness as the lakes and surrounding ecosystems are now restored to their natural, fishless state. It minimizes ecological manipulation of wilderness for human ends, and places primary value on restoring ecological integrity in wilderness. It sacrifices some recreational and angling opportunities at these 11 wilderness lakes, but allows for restocking WCT in the many lakes in the South Fork drainage outside the BMWA.

Desert Bighorn Sheep Management in the Kofa Wilderness, Arizona

The second case study we examined as a class also involves a potentially endangered species, but rather than a one-time short-term "trammeling" of wilderness to remove nonnative species to benefit native species, this case contemplates permanent and ongoing manipulation of a wilderness to benefit one of its resident native species. Because some of the elements of the wilderness management dilemma in this case are similar to those of the Bob Marshall Wilderness case just discussed, I move quickly through those aspects and focus instead on what is distinctive to the case and how the students proposed to resolve it.

The case here involves preserving populations of desert bighorn sheep within the Kofa National Wildlife Refuge near Yuma, Arizona. Highly valued as both a trophy hunting animal that brings in a significant amount of revenue to state agencies through the sale of hunting licenses, as well as an iconic species that symbolizes wildness in the desert southwest, the ongoing viability of the desert bighorn is threatened throughout the southwest by a combination of habitat fragmentation from highways and residential construction, and ongoing severe drought—together drought and habitat loss make it difficult if not impossible for desert bighorn to migrate between isolated mountain ranges to find water.[25]

Hence, in 2007 the US Fish and Wildlife Service (USFWS) authorized the construction of permanent water tanks—or "water guzzlers"—within the Kofa Wilderness to provide year-round sources of water for the sheep. Not only the construction, but also the ongoing maintenance of the guzzlers requires regular access by motorized vehicles within the wilderness. Predictably, wilderness advocates immediately contested this decision in court. One such group, *Wilderness Watch*, questioned whether the USFWS "may construct, maintain, and operate permanent water impoundments inside a federally designated wilderness area, when the Wilderness Act ... explicitly prohibits the use of motorized vehicles, temporary roads, and permanent structures in wilderness."[26] The USFWS responded that the Wilderness Act does not prohibit all permanent structures and that Congress recognized that some structures might be necessary for administering a wilderness area. To the dismay of the plaintiff groups, the Arizona District Court ruled in favor of the USFWS on all counts[27]—thus resolving for now the *legal* issues of the case, but leaving unresolved the underlying *ethical* dilemma:

> Unless provided with year-round sources of water, the desert bighorn sheep of the Kofa Wilderness may experience increased mortality

> But

> Providing water for bighorn sheep in Kofa will adversely affect both the *untrammeled state* and the *natural condition* of the Kofa wilderness.

Working through step one of the framework—Informational Scoping—reveals a good deal of complex and relevant background information that I can only allude to here, including the history of the Kofa Wilderness Area and the reasons for its establishment, biological information on the desert bighorn sheep and

---

[25]   Arizona Game and Fish Department (AGFD) and Kofa National Wildlife Refuge. 2007a. *Investigative Report and Recommendations for the Kofa Bighorn Sheep Herd.* Prepared by USFWS and AFGD; B. Campbell and R. Remington, "Influence of Construction Activities on Water-Use Patterns of Desert Bighorn Sheep," *Wildlife Society Bulletin* 9 (1981): pp. 63–5; H. E. McCutchen, "Desert Bighorn Sheep," in T. J. Stohlgren (ed.), *Our Living Resources: A Report to the Nation on the Distribution, Abundance, and Health of U.S. Plants, Animals, and Ecosystems* (Washington, DC: U.S. Geological Survey, 1995).

[26]   *Wilderness Watch et al. v. U.S.FWS, et al.* 2007.

[27]   Arizona Game and Fish Department (AGFD). 2007b. "District Court Ruling lets Stand Wildlife Water Development on Kofa Wildlife Refuge." Retrieved February 16, 2009 from www.azgfd.net/wildlife/conservation-news/district-court-ruling-lets-stand-wildlife-water-development-on-kofa-wildlife-refuge/2008/09/15/.

their reproduction and mortality rates, the influence of the international border with Mexico and the effects of efforts to curb illegal immigration on the area's ecosystems, the historical use and knowledge of desert bighorn sheep by Native Americans, and debates about the reasons and causes for current declines in sheep populations. In addition, although there remains much uncertainty regarding the success of watering structures in maintaining sheep populations, some research has indicated that where water is maintained in areas similar to Kofa with resident sheep populations, those populations are fairly stable in comparison with Kofa.[28] Hence the Arizona Game and Fish Department (AGFD) management strategy to provide year-round water sources, despite apparent contradictions with the Wilderness Act. Yet as the graduate students point out, "providing sheep with water—especially through construction of invasive impoundments—diminishes wildness of the sheep and has unknown consequences for the ecosystem of which they are a part. Additionally, construction of permanent water impoundments is contrary to the congressionally mandated designation that wilderness areas are to remain 'untrammeled' and in their 'natural condition'."

As in the case with restoring native fish in the Bob Marshall Wilderness Area, briefly reviewing the core ethical commitments of the model reveals some of the complexity of the case. In terms of *commitment to wilderness integrity*, several interesting questions arise for the wilderness managers, such as:

- What is considered "appropriate use" of the Kofa wilderness?
- Should preservation of bighorn sheep populations take precedent over preservation of other wilderness components?
- When they conflict, which wilderness value should be given higher priority: untrammeledness or naturalness?
- Should hunting and other sources of revenue play a role in the decision-making process for Kofa Wilderness despite the fact that the intent of the Wilderness Act was to protect such lands from human influence and commercialization?

The *commitment to participatory process* was largely circumvented by the government agencies involved when they sought a "categorical exclusion" under NEPA that exempted them from soliciting public comment. This, however, did not prevent several groups from joining together in a lawsuit to prevent the agency actions. *Commitment to foresight as informed by the Precautionary Principle* suggests that the burden of proof lies with the government agencies proposing the action. In this case where taking no action would likely lead to a decline in sheep population and thus affect the naturalness of the area due to anthropogenic influences, commitment to the precautionary principle can be seen both in taking *some* action, but carefully scrutinizing any action taken.

---

[28]   AGFD and Kofa National Wildlife Refuge, *Investigative Report*.

The students' research demonstrated that attention to *traditional ecological knowledge and tribal sovereignty* had a great deal to contribute to resolving this case, though neither was consulted by the government agencies. Because of their extensive history with both the landscape and with desert bighorn sheep, local tribes have a great deal of ecological knowledge and a high stake in how this case is resolved. Finally, the specific management mandates of the AGFD and USFWS must be harmonized with those of the Wilderness Act, such as AGFD's responsibility in "Conserving, enhancing, and restoring Arizona's wildlife resources and habitats through protection and management programs, and providing wildlife resources for the enjoyment, appreciation, and use of people."[29]

In their research, the students identified six initial options, including the agency's preferred option. These include:

> Option 1: no action (indefinitely): no structures in wilderness (though structures *outside* wilderness may be permitted).
> Option 2: no action (temporarily—at least five years) to allow for more studies.
> Option 3: focus on other potential causes of sheep decline.
> Option 4: only allow for enhancements to *natural* structures.
> Option 5: build structures, but reduce motorized access thereafter.
> Option 6: install underground water structures and allow motorized access (agencies' preferred action).

Space permitting, it would be instructive to examine each of these options in terms of the core ethical commitments, and the likely effects on both maintaining sheep populations and maintaining wilderness values in management. Each has significant pros and cons, and it quickly becomes apparent that no single option can avoid either risk to the sheep or compromising wilderness. The students noted, however, that Option 6, the agencies' preferred option, violates several of the core ethical commitments of the model:

> It encourages neither the "untrammeled" nor "natural" values of wilderness. Construction of permanent, large-scale water impoundments trammels wilderness. New water sources impact naturalness in two key ways. Guzzlers enlarge mountain lion habitat and diminish bighorn "escape habitat," increasing predation. Guzzlers may change bighorn foraging and water seeking behavior, increasing the fitness of less drought-adapted sheep. These outcomes would also constitute hobbling of the wilderness ecosystem's self-sustaining character.

---

[29]   Arizona Game and Fish Department, *Wildlife 2006: The Arizona Game and Fish Department's Wildlife Management Program Strategic Plan for the Years 2001—2006.* 2006. Retrieved February 16, 2009 from www.azgfd.gov.

And, of course, adding a significant component such as water to a desert ecosystem is bound to have all sorts of unanticipated effects and consequences for the many other members of the ecosystem, both flora and fauna. Similarly, Option 1, the take no action option, is likely unacceptable because it violates the Wilderness Act's mandate to protect naturalness, and because of the high value attached to desert bighorn sheep in desert wilderness.

To begin to generate possible alternatives that might be more acceptable to a majority of stakeholders in this case, the students generated a list of commonalities and conflicts among stakeholders that should be brought into consideration. These include:

1. All stakeholders agree that humans should play some role in managing ecosystems; no group views humans as outside of nature. A conflict arises between the US FWS placement of humans as obligated to reparations toward nature, and tribal and wilderness advocates' placement of humans as crucially part of nature but obligated to foresight and restraint.
2. Naturalness and untrammeledness often are in conflict in this case.
3. Trammeling of wilderness constitutes an unacceptable risk to some stakeholders.
4. Bighorn extinction is unacceptable to all parties, so only a low risk to sheep is acceptable. Bighorn domestication and lion hunting are unacceptable to wilderness advocates, lion scientists and tribes. Risk of extinction is difficult to determine because of inconclusive scientific evidence.

In addition, part of the ethic of virtually all stakeholders is the conviction that bighorn sheep should be maintained as a wild, natural species, as wild bighorns epitomize wilderness. To do this, however, means addressing the wildness of the bighorn's *habitat*.

With this in mind, the students revisited the different options to come up with two new options that represent potential compromises that maximize satisfaction of core commitments to each stakeholder group. They concluded, "The resolution will strike a balance between moral obligations for intervention in, and disengagement from, natural processes. It will also strive to achieve dynamic equilibrium among different stakeholder values including hunting, recreational and spiritual land use, and cultural views of water. The resolution could be stated as follows: *Preserving sheep populations that retain wild characteristics requires maintaining core habitat areas where hydrologic cycles are free from human interference* and *maintaining permanent water sources where such sources do not degrade the ecosystem as a whole.*"

Here are their descriptions:

Option 7: a land-based compromise

This option would delineate spatial zones where different ethics apply and would specify appropriate adaptive management plans within the different zones. It would designate "backcountry" areas where no trammeling would occur, "mid-country" areas where low-impact construction would be permitted, and "front country" zones where high-impact construction and maintenance projects would be permitted. It would allow managers to manage high-quality bighorn habitat outside of wilderness as "backcountry" without redrawing wilderness boundaries.[30]

Option 8: a time-based compromise

This option would limit new construction but permit maintenance (guzzler refilling) at certain times of year. It would use seasonal closures to minimize human disturbance during key times of the year such as breeding and lambing seasons, and would ensure tribal use of the land for hunting or other purposes at culturally significant times of the year. Maintenance could coincide with natural ecosystem cycles by, for example, refilling *tinajas* before sheep seek them out. Fragmenting use over time encourages *adaptive* management in which agencies respond to natural cycles in the system and only trammel it when absolutely necessary (during prolonged drought, for example). Rather than designating zones where water is always present, a more time-based approach could establish "triggers" that only set off the impact when certain extreme conditions are in place. Land boundaries are extremely permanent and contentious in American society. Time boundaries are more likely to be flexible and adapt to changing conditions in the future.

Concluding the ethical analysis, either Option 7 or 8 could be seen as the "best" option for the agency to take because neither violates core commitments

---

[30] Backcountry zones would ensure preservation of the wilderness view-shed, the wildness of wildlife, and the culturally significant tribal sites, while also providing the scientific community with long-term ecological research areas. Backcountry areas could incorporate habitat areas high in barrel cactus or other non-tank water sources, thus limiting guzzler construction that might diminish forage quality.

In mid-country zones, limited enhancement of natural water impoundments would be permitted. Enhancement might be based on indigenous water harvesting practices or may be accomplished using local materials and non-motorized tools. Many natural impoundments are ephemeral and would dry up seasonally or during prolonged drought if they are only minimally altered, so these modified structures would maintain historical fidelity and "natural" bighorn water-seeking behavior. This zone could serve multiple purposes: preserving culturally-significant tribal sites, allowing visitors to encounter solitude and minimal evidence of human intervention, and serving as longtime ecological research sites and controls in scientific experiments (that investigate, for example, guzzler efficacy).

In the front country, human-made impoundment structures could be permitted in zones where it can be reasonably demonstrated that they: (1) will not have, or have the least, major detrimental impacts on the greater ecosystem's hydrologic cycle (i.e. lower in the watershed where flood volumes are greater); (2) there are no adjacent suitable areas outside the wilderness boundary; and (3) guzzlers will not compromise the ability of other zones to provide for all other stakeholders' values, including, for example, "opportunities for solitude," and traditional spiritual/hunting uses. Agencies could prioritize acquisition of additional front country beyond the refuge boundary to improve sheep access to existing water sources.

and both uphold the unifying ethics to a relatively high degree, while also allowing for conflicting ethics to be satisfied. They limit human intervention using different paradigms—space and time—but both maintain the sheep as wild, natural species in a way that causes no further degradation of ecosystem integrity (primarily free from human influence)—at least in certain "zones" or during certain times of the year.

Conclusion

As human-induced pressures on ecological integrity increase through such factors as climate change, habitat loss and invasive species, so too will the opportunities for and pressures to carry out ecological restoration in wilderness. Yet such actions constitute a dilemma for wilderness managers and the interested public, as restoration necessarily entails the deliberate manipulation of ecosystems and ecological processes—even if only short-term—that violate the spirit and perhaps the law as embedded in the Wilderness Act. Complicating these actions further is the diverse array of stakeholders in wilderness, each with its own set of values and commitments for what should—or should not—be done in wilderness. An ethical model designed to engage multiple stakeholders in deliberation together about the *ethical* dimensions of proposed restoration in wilderness may help to (1) identify core ethical commitments; (2) generate and assess several options; and (3) seek an option that best meets the range of stakeholder values and commitments. When carried out by a stakeholder group over time, it may assure the best ethical deliberation about particular wilderness areas by those who know it best and are committed to its long-term well-being. It can facilitate thoughtful and constructive public engagement with agency decision-making and reduce or avoid costly litigation over agency management decisions.

Acknowledgments

The author gratefully acknowledges the following students who participated in the Fall 2008 class, "Ethical Issues in Ecological Restoration" that developed the framework for ethical analysis and the two case studies reviewed in this chapter: Talasi Brooks, Whitney Byrd, David Colbeck, Oskar Cole, Laurel Douglas, Amy Edgerton, Erika Edgely, Katie Makarowski, Larry McKay, Yvonne Sorovacu, Bethany Taylor, Emily Thompson, Cleo Woelfe-Erskine. Oskar Cole had primary authorship of the section presenting and discussing the framework. David Colbeck had primary authorship of the Bob Marshall case study. Cleo Woelfe-Erskine, Amy Edgerton and Katie Makarowski had primary authorship of the Kofa Wilderness case study. Peter Landres read and commented on the entire manuscript and made several constructive comments. Any errors or inaccuracies are the responsibility of the author.

Bibliography

Arizona Game and Fish Department (AGFD). 2006. *Wildlife 2006: The Arizona Game and Fish Department's Wildlife Management Program Strategic Plan for the Years 2001—2006*. 91 pp. Retrieved February 16, 2009 from www.azgfd.gov/.

Arizona Game and Fish Department (AGFD) and Kofa National Wildlife Refuge. 2007a. *Investigative Report and Recommendations for the Kofa Bighorn Sheep Herd*. Prepared by USFWS and AFGD. 39 pp.

Arizona Game and Fish Department (AGFD). 2007b. "District Court Ruling lets Stand Wildlife Water Development on Kofa Wildlife Refuge." Retrieved February 16, 2009 from www.azgfd.net/wildlife/conservation-news/district-court-ruling-lets-stand-wildlife-water-development-on-kofa-wildlife-refuge/2008/09/15/.

Bonneville Power Administration. 2004. South Fork Flathead Watershed Westslope Cutthroat Trout Conservation Program, Draft Environmental Impact Statement (DOE/EIS-0353). Retrieved February 7, 2009 from www.gc.energy.gov/NEPA/draft-eis0353.htm.

Bonneville Power Administration. 2005. South Fork Flathead Watershed Westslope Cutthroat Trout Conservation Program, Final Environmental Impact Statement (DOE/EIS-0353). Retrieved February 7, 2009 from www.gc.energy.gov/NEPA/draft-eis0353.htm.

Bonneville Power Administration. 2006. South Fork Flathead Watershed Westslope Cutthroat Trout Conservation Program, Record of Decision. Retrieved February 7, 2009 from http://regulations.vlex.com/vid/flathead-westslope-cutthroat-trout-22139807.

Borgmann, A., *Technology and the Character of Contemporary Life* (Chicago: University of Chicago Press, 1984).

Callicott, J. B. and M. P. Nelson, *The Great New Wilderness Debate* (Athens, GA: University of Georgia Press, 1998).

Campbell, B., and R. Remington, "Influence of Construction Activities on Water-Use Patterns of Desert Bighorn Sheep," *Wildlife Society Bulletin* 9 (1981): pp. 63–5.

Cole, D. N., "Management Dilemmas that will Shape Wilderness in the 21st Century," *Journal of Forestry* 99 (2001): pp. 4–8.

Cole, D. N. and W. E. Hammit, "Wilderness Management Dilemmas: Fertile Ground for Wilderness Management Research," in D. N. Cole, S. F. McCool, W. A. Freimund, and J. O'Loughlin (eds), *Wilderness Science in a Time of Change Conference—Volume 1: Changing Perspectives and Future Directions; 1999 May 23–27; Missoula, MT. Proceedings RMRS-P-15-VOL-1* (Ogden, UT: U.S. Department of Agriculture, Forest Service, Rocky Mountain Research Station, 2000).

Cronon, W., *Uncommon Ground: Rethinking the Human Place in Nature* (New York: W. W. Norton, 1996).

Gruber, J. S., "Key Principles of Community-Based Natural Resource Management: A Synthesis and Interpretation of Identified Effective Approaches for Managing the Commons," *Environmental Management* 45 (2010): pp. 52–66.

Hamel, R., "A Process for Ethical Decision Making," Personal communication with author by email, September 22, 2008.

Higgs, E. S., "What is Good Ecological Restoration?" *Conservation Biology* 11 (1997): pp. 338–48.

Higgs, E. S., *Nature by Design: People, Natural Process, and Ecological Restoration* (Cambridge, MA: MIT Press, 2003).

Higgs, E. S., "Restoration Goes Wild: A Reply to Throop and Purdom," *Restoration Ecology* 14 (2006): pp. 500–3.

Higgs, E. S. and R. Hobbs, "Wild Design: Intention, Intervention, Reciprocity" in D. Cole and L. Yung (eds), *Conserving Parks and Wilderness in the 21st Century: Naturalness and Beyond* (Washington, DC: Island Press, 2009).

Landres, P., C. Barns, J. Dennis, T. Devine, P. Geissler, C. McCasland, L. Merigliano, J. Seastrand, and R. Swain, "Keeping it Wild: An Interagency Strategy to Monitor Trends in Wilderness Character across the National Wilderness Preservation System," Gen. Tech. Rep. RMRS-GTR-212. (Fort Collins, CO: U.S. Department of Agriculture, Forest Service, Rocky Mountain Research Station, 2008).

Landres, P. B., S. Boutcher, L. Merigliano, C. Barns, D. Davis, T. Hall, S. Henry, B. Hunter, P. Janiga, M. Laker, A. McPherson, D. S. Powell, M. Rowan, and S. Sater, "Monitoring Selected Conditions Related to Wilderness Character: A National Framework," Gen. Tech. Rep. RMRS-GTR-151 (Fort Collins, CO: U.S. Department of Agriculture, Forest Service, Rocky Mountain Research Station, 2005).

Landres, P. B., M. W. Brunson, and L. Merigliano, "Naturalness and Wildness: The Dilemma and Irony of Ecological Restoration in Wilderness," *Wild Earth* 10 (2001): pp. 77–82.

Landres P. B., R. L. Knight, S. T. A. Pickett, and M. L. Candenaso, "Ecological Impacts of Administrative Boundaries," in *Stewardship Across Boundaries* (Washington, DC: Island Press, 1998), pp. 39–64.

Landres P. B., P. W. White, G. Aplet, and A. Zimmermann, "Naturalness and Natural Variability: Definitions, Concepts, and Strategies for Wilderness Management," in D. L. Kulhavy and M. H. Legg (eds), *Wilderness and Natural Areas in Eastern North America: Research, Management, and Planning* (Nacoghoches, TX: Center for Applied Studies in Forestry, Stephen F. Austin State University, 1998a), pp. 41–50.

McCutchen, H. E., "Desert Bighorn Sheep," in T. J. Stohlgren (ed.), *Our Living Resources: A Report to the Nation on the Distribution, Abundance, and Health of U.S. Plants, Animals, and Ecosystems* (Washington, DC: U.S. Geological Survey, 1995).

Margles, S. W., R. B. Peterson, J. Ervin, and B. A. Kaplin, "Conservation Without Borders: Building Communication and Action Across Disciplinary Boundaries for Effective Conservation," *Environmental Management* 45 (2010): pp. 1–4.

Muñoz-Erickson, T. A., B. Aguilar-González, M. R. R. Loeser, and T. D. Sisk, "A Framework to Evaluate Ecological and Social Outcomes of Collaborative Management: Lessons from Implementation with a Northern Arizona Collaborative Group," *Environmental Management* 45 (2010): pp. 132–44.

Raffensperger, C. and J. Tickner, *Protecting Public Health and the Environment: Implementing the Precautionary Principle* (Washington, DC: Island Press, 1999).

South Fork Flathead Watershed Westslope Cutthroat Trout Conservation Program Q&A Fact Sheet: June 8, 2004. Retrieved February 7, 2009 from http://fwp. mt.gov/content/getItem.aspx?id=10435.

Throop, W. and R. Purdom, "Wilderness Restoration: The Paradox of Public Participation," *Restoration Ecology* 14 (2006): pp. 493–9.

Walton, A. A., "Conservation Through Different Lenses: Reflection, Responsibility, and the Politics of Participation in Conservation Advocacy," *Environmental Management* 45 (2010): pp. 19–25.

Watson, A., L. Alessa, and B. Glaspell, "The Relationship between Traditional Ecological Knowledge, Evolving Cultures, and Wilderness Protection in the Circumpolar North," *Conservation Ecology* 8 (2003): p. 2.

Wilderness Act of 1964 (16 U.S.C. 1131-1136, 78 Stat. 890) Public Law 88-577, approved September 3, 1964.

Wilderness Watch et al., Plaintiffs, v. *U.S. Fish and Wildlife Service et al.*, Defendants. U.S. District Court for the District of Arizona. 2007. Plaintiffs' Motion for Summary Judgment and Accompanying Memorandum of Points and Authorities. No. CIV-07-1185-PHX-MHM. Case 2:07-cv-01185-MHM Doc. 78.

Zahniser, H., "Editorial: Guardians not Gardeners," *The Living Wilderness* 83 (1963): p. 2.

Chapter 5

Re-creation:
Phenomenology and Guerrilla Gardening

Mélanie Walton

On a sadly tenacious street corner in a well-worn Midwestern town, my attention had been arrested; I was frozen, despite the threat of rain from the dreary spring sky. My shoes were dusty, having just cut across an abandoned lot full of gravel, dumped sand, and broken beer bottles. Captivated, I was standing before a patch of blooming daffodils. There was neither soil nor garden, just rubble where the gravel lot gave way to a broken sidewalk by a ragged street of endless cars in a hurry to or from somewhere. No one else seemed to notice, let alone care, about the unlikely patch of daffodils. *The world gives itself to us only insofar as we give ourselves to the world.*

This dictate is the most basic premise of phenomenology, a widely influential school of contemporary continental philosophy. Phenomenology is a primarily epistemological theory that captures how an abstracted event, like an encounter with the unexpected, gives one pause and permits a flood of visions, each an aspect of experience commonly overlooked. Its result can be a vision of an ethical ideal previously unthought, while its repeated use as an epistemic method can then pragmatically and profitably guide one in the betterment of the human-battered world. The unlikely horticultural display at my local intersection, the daffodils' radical reclamation of its rubble, phenomenologically considered, reveals how guerrilla gardening, an illicit recreation of the urban grey to green, can be environmental phenomenology's preeminent example. Phenomenology demands the suspension of biased thought, including basic subject–object relationships that may set humanity against nature, and encourages thinking through multiple possibilities by the experiential opening of oneself to usually unacknowledged modes of relation. Appreciating guerilla gardening as more than acts of trespass and vandalism with seed or bulbs necessitates the traits taught by phenomenology. These intellectual and horticultural skills coalesce into the most fruitful method of environmental recreation, that is, a practical, meaningful way to think through and transform sites of urban decay into those for the sustenance and flourishing of both human activity and nature.

Despite the seeming vacancy of the gravel lot, the everyday world is not a blank canvas for humanity to inscribe, but an ecosystem already created and continually in flux, one whose constancy is only in the openly definable interconnections betwixt it and humanity. The clump of daffodils illuminated, for me, that in order to define how we create or place or remove our mark on the world, we must seek to understand these interconnections by being open to experience and able to describe them—in which phenomenology preeminently aids us. Before exploring how guerrilla gardening's intentional creation of sites that, in their poignant juxtaposition of the urban and natural, can phenomenologically initiate reflection and further inspire us to will our neglect into positive action, to reclaim verdure out of our environmental damage, we must understand the basic phenomenological process of experience and how its theory transforms into ecologically sound praxis.

Phenomenology's seminal refrain, that the world gives itself to us as we give ourselves to the world, indicates a mode of experience wherein our presumptions of an egoistic supremacy over an essentially independent world (in which we are its sole interpreter and judge of its value) are unsettled. This suspension of bias, called the *epoché*, reveals our relation with the world to be more reciprocal than we regularly presume, letting us see, more clearly, what is essential within this interaction and its dynamism. The task of this theory of knowledge, founded in the 1920s by the German thinker Edmund Husserl, is primarily methodological: to let that which shows itself be seen from itself in the very way in which it shows itself. Husserl instructs us to attend to the things themselves and we honor his maxim by engaging the fluidity of perspectives of experience without bias. This method reveals the essential meaning of the world by laying bare the nature of experience— that is, meaning is founded by our interaction with the world. According to an etymological interpretation of phenomenology, through *phainomenon* and *logos*, by Husserl's student Martin Heidegger, the things themselves, the essential nature of the world and its contents, present a "seeming" that requires a "seeing" and "discourse."[1] In other words, the world *seems*, thus requiring a subject to see it, listen to it, and participate in its appearance and elaboration, which forms its meaning. Thus, phenomenology is a methodological study of everything that shows itself and explicates itself to us *when* we turn our focused gaze upon it and attuned ear to it—which, it reminds us, we do not often do in the hasty course of daily life.

Phenomenology's guiding expression thus epitomizes how this school of thought can become a working praxis for the (literally) dirty work needed to transform human-wrought dilapidation back into something lushly thriving, humanly and environmentally beneficial. Its initiation requires us to give ourselves to the mutually giving world; it requires a shift of perspective, a suspension of bias, to shake us from our egoism that takes the world for granted, independent from us and static. The event of an unexpected appearance, like bright daffodils amidst rubble, can jar us from our blindly presumptive everydayness, show us what it is, hearken to how it was, indicate what could be, and, when there is an

---

[1]   Martin Heidegger, *Being and Time*, trans. Joan Stambaugh (New York: SUNY Press, 1996), Introduction II, §7, pp. 23–34.

uncomfortable disjuncture between reality and the ideal, inspire us to act.[2] Being primarily epistemological, phenomenology seeks to yield essential descriptions, a better knowledge of the interrelationships of the world and humanity. But this epistemology is not a removed, objective study; it is the knowledge of *lived* experience, of an embedded, aware subject actively engaged with her environment and, thus, I argue, it is ethical and inspirationally pragmatic. Its demand for the attentive focus of our gaze cultivates multi-dimensional observation that inspires us to see potential while suspending preconceived ideas, for example, where daffodils belong, the definition of a garden, and the helpless state of urban decay. Its resultant contemplation gives us an ethical goal and reveals the steps along a path from reality to the attainment of that ideal.

Before further exploring the environmentally perfect fusion of guerrilla gardening and phenomenology (wherein the former can both jumpstart the latter and be its obvious conclusion), it is necessary to further explore how phenomenology can sketch a theoretical guidebook for green action. The desired end of its instructed attention to the things themselves is to be able to give a direct description of our experience as it is in itself, born from our interrelation with the world, without our accumulated biases from historical, psychological, or scientific modes of thinking. These perspectives are not in themselves wrong, nor can they correctly be entirely left out of knowledge. Instead, their temporary phenomenological suspension reveals the nature of meaning to be more complex and dynamic than they can capture and how each of these limited perspectives must be acknowledged as only single modes among many by which to experience and describe the multi-dimensional meaning of the world. This insight aids an environmental plan by complicating the object of its study and, thereby, necessitating a balance of approaches.

In the *Crisis of the European Sciences*, Husserl elaborates this idea of a complicated world as the *Lebenswelt*, the life-world or the world of lived experience, in contrast to a world constituted by a mathematization of nature.[3] Galileo geometrized nature as Descartes and Leibniz arithmetized it wherein, according to the scholar David Carr's reading, "nature becomes a mathematical manifold and mathematical techniques provide the key to its inner workings."[4] Science, Husserl argued, sought to find in nature a precision and uniformity that is lacking in its everyday experience. This exactness is achieved, first, by abstraction

---

[2] For an intensive analysis of how a shift in perspective can provoke something both emotive and reflective, which I argue can then initiate action, see David E. Cooper, "Phenomenology and 'Atmosphere'," in *A Philosophy of Gardens* (Oxford: Clarendon Press, 2006), pp. 47–53.

[3] Edmund Husserl, *Crisis of the European Sciences*, trans. David Carr (Evanston: Northwestern University Press, 1970), pp. 23ff.

[4] David Carr, "Husserl's Problematic Concept of the Life-World," in *Husserl: Expositions and Appraisals*, ed. Frederick Elliston and Peter McCormick (Notre Dame, IL: University of Notre Dame Press, 1977), pp. 202–12, p. 205.

from the appearances given to perception and, then, by their interpretation. The result is that science *imposes on* nature a precision foreign to any actual experience of it. Further gilding this faux-precision, Husserl argues, science imposes an ontological claim, *to be is to be measurable*, and then extrapolates, as consequences, various further epistemological claims, ultimately abstracting nature further and further from actual dirt, rubble, and daffodils.[5] Science that begins from such a mathematized foundation, bases all its subsequent theories of knowledge upon an abstraction from reality that presumes the world to be uniform and predictable. Strains of contemporary science reject these presumptions, but the simple encounter with daffodils growing amidst urban rubble on a neglected street corner plainly shows the experience of nature to be of events neither uniform nor predictable, but certainly full of meaning that extends far beyond the empirically measurable.

Phenomenology is, of course, also an abstraction; indeed, its central most important insight is a *method* of abstraction: the *epoché*. But it is acutely aware of its nature and avoids its traps. That is, precisely because it employs a *method* of abstraction, it cannot overlook its origin in and the meaningfulness of the *Umwelt*, the world that is intuitively given and from which one can then abstract all the secondary qualities of appearance. Husserl names this the "pre-scientific life-world" and, instead of fearing the elusive and vague, non-uniform appearances, it is a perspective that seeks to take in the vagaries and be moved by them.[6] All that is, is a nebulous flux of appearances, but it is also a *world*; grasping the cohesion of difference is to give it meaning. Our attention to the immediately given reveals that the world is not an abstraction, a formula, or a mere, static mental image of nature. Instead, it shows us that the mathematized world of the early moderns is just one perspective of the world amongst many, from merely one attitude, one mode from the plethora for perception.

In addition to the mathematical attitude, Husserl differentiates two further, more fundamental modes of perception: the natural attitude and the phenomenological attitude. Within the natural attitude, one is within the everyday, automatically directed through pre-given frameworks of intentional relations by which one *naturally* accords independent existence to the correlates of consciousness or things intended. That is, I take myself to be the central perspective, but also a mere actor in a staged world where the set has meaning, even without me around. For example, I purposefully cut across the vacant lot; I am the subject of this world, taking it on presumption that it is likely owned by someone, even while carelessly piled with gravel and sand that dust over my shoes as I sidestep the glittering shards of broken bottles and swerve to avoid the small, seemingly misplaced birds with long legs frolicking in the powdery dirt. In the natural attitude, I take each element of this view to have independent, objective existence and predetermined meaning. In contrast, in the phenomenological attitude, one redirects one's

---

5   Husserl, *Crisis*, pp. 51ff.

6   *Ibid.*, p. 43.

attention from the objects themselves to the intentional relation in and through which one posits the objects as such. In this attitude, I reconsider my relation to and with the environment, my presumptions about ownership and neglect, the origins of the bottles, and the misplacement of the birds.

Husserl's work *Ideas I* elaborates this shift by delineating the many dimensions of reflection in each attitude through the experience of a blossoming apple tree. His example illustrates the transformation that consciousness undergoes concerning experience, intentionality, and reflection as a result of adopting the transcendental stance of the phenomenological attitude. The experience of the tree in full flower, in the natural attitude, is as *Erfahrung*: the viewer experiences a presentation of the tree as a substance possessing properties of one kind or another—it is a Macintosh, an old, established one, healthy and abundant. In the phenomenological attitude, experience becomes *Erlebnis*, lived experience: this is a dynamic encounter wherein the ego is engaged in meaning formation—the tree moves me, its beauty affects me—as opposed to passively receiving a static, predetermined presentation of the tree. Intentionality, in the natural attitude, is presented as if the ego posits the tree as existing entirely distinctly and independently of the consciousness. In the phenomenological attitude, the positional consciousness becomes *explicitly* intentional, wherein it is not the object itself so much as it is the intentional relation *to* the object that now becomes the focus of attention. Finally, reflection, in the natural attitude, intervenes only as a specific redirection of attention from the object to the subject—I pause my deduction that its cultivar is Macintosh to think about whether or not I like Macintosh apples. In the phenomenological attitude, in contrast, one becomes aware that all consciousness is implicitly reflective in character—I am instantly and intuitively aware of the reflexive, meaningful relation between the tree and me.[7]

In other words, phenomenology provokes a concise shifting of my attention from the rubble, beer bottles, and birds to my interrelation with them on these many levels. This shift transforms reflection into an awareness of how relation is an implicitly interwoven, yet overlooked, feature of all consciousness. And, while to know anything about these things around me I invoke my past experiences, phenomenology's *epoché* permits me to come to know them more essentially without bias. I no longer think of the birds as sandpipers, as misplaced, the lot as vacant or owned, the gravel and bottles as breaking laws, but think, instead, of how I give myself to the lot and how the lot gives itself to me. Phenomenologically, it spurs ideas of barrenness, decay, and dilapidation; it made me dusty. My natural interrelation with it was one of ignorance; as gray sky melded with gray ground to eliminate all spark, passion, life, and color from the surroundings, I ignored its existence until the patch of the daffodils at the lot's far edge re-colored my experience. The unexpected event was what called me from the natural to the

---

7 Edmund Husserl, *Ideas Pertaining to a Pure Phenomenology and to a Phenomenological Philosophy, First Book*, trans. F. Kersten (The Hague: Martinus Nijhoff Publishers, 1983), §88.

phenomenological attitude; it made the lot essentially come into existence. It prompted my awareness of my interrelation with my surroundings and offered me a wholly new perspective on the lot and a whole lot more. In their almost absurd superfluity, the daffodils told me that the dilapidated space could be otherwise. An ecological ideal and an initiative to action, both ethical and pragmatic, naturally followed from the epistemological exercise.

The most practical application of phenomenology to environmental restoration or recreation is as an explicit exercise of the phenomenological attitude's suspension of prejudice and recognition of the perspectival nature of perception, which acknowledges the many actual and potential instantiations of any given environment. David Utsler's contribution within the current volume (Chapter 10) explores the hermeneutical dimension of this idea in his elaboration of "environmental identity" and likewise supports the ethical implications that issue from our related theoretical groundings. Akin to how hermeneutics and phenomenology redirect thought back to the generation of meaning, environmental restoration suggests that there is an original model from which a place has diverged and calls to be redirected back. Once one becomes aware and capable of defining a place's originary state through rereading or the phenomenological *epoché*, environmental restoration calls for the implementation of a scientific model, that is, an alteration between the mathematized and natural attitudes, to uncover the correct path of return. To environmentally recreate a place equally invokes phenomenology, yet utilizes it to uncover an end and its path that are not necessarily pre-given. Recreation suggests a transformation that defies the idea of a return to an original mode of being. An urban wasteland, for example, calls for environmentally sensitive recreation when neither a return to wilderness nor a new creation of city concrete makes sense. It, then, calls to us for assistance in becoming something other than what it is or was without, providing us an end picture and road map. When recreation lacks laws to follow, it is best guided by phenomenology.

For recreation, phenomenology's revelation of our lived experience of the world can permit the emotive to take precedence over the only-rational comprehension of reality and encourage thinking far outside of the grid. Phenomenology's methodological turning of child-like eyes to the world, this seeking of a descriptive knowledge that can capture the awe of embedded awareness, is a battle-ready weapon against close-minded thinking and is a necessity for successful recreation. While restoration's definable goals achievable by inscribed methods allow it to have more clearly distinguishable proponents and opponents, recreation, lacking a single indubitable end and path, lacks such clear-cut friends and enemies. Thus, phenomenology's rigorous affinity with thinking plurality and difference, embracing the elusive and encouraging the unexpected, makes it a superior theory to employ in environmental action. Invoked by my own attention's arrest by an odd patch of daffodils and subsequent flourish of phenomenological reflections, one example of recreation that holds great promise for positive urban environmental change is guerrilla gardening.

According to Winston Churchill, war and gardening are the two most natural preoccupations and pursuits of humanity.[8] Guerrilla gardening is their synthesis; it is a horticultural practice that evades the law and any standard definitions of a garden and its activity. According to Richard Reynolds' book *On Guerrilla Gardening*, war and gardening are one because, "in both exploits you wrestle forces beyond your control, you shape the landscape and you get messy ... Both ... are creative as well as destructive."[9] David Tracey's "manualfesto," *Guerrilla Gardening*, warns, "it is not for everyone ... It's not all sunshine and daisies ... [it] may sound romantic, but the work can be grueling. Sites may, by definition, be uninviting. Growing conditions are often harsh. Routine chores can be a logistical grind," not to mention that "You might get arrested."[10] While its arrests are yet undocumented, the *Los Angeles Times* has noted guerrilla gardening to be "a burgeoning movement of green enthusiasts who plant without approval on land that's not theirs ... free-range tillers [who] are sowing a new kind of flower power."[11] However, this green empowerment is not exactly new: it harmonizes with the Johnny Appleseed activism, most political cases of agricultural squatters, and even the fervor behind the legal, worldwide wartime "victory gardens." While environmental obligation and the phenomenologically productive capture of awe are seemingly at odds with the excessively violent imagery invoked by comparing gardening to guerrilla warfare, the correlation does help to reveal the very complexity of meaning embedded in a garden and its environmentally recreative activity.

When I encountered a hearty clump of daffodils in litter-strewn urban decay, my attention was struck by their beauty and existence amidst adversity. Their instant beauty compelled me to attend to them in and of themselves, which I could only do by noting their contrast to the vacant lot and, thus, rethinking their interrelation with me in order to consider our cooperatively created meaning. Their appearance challenged the everyday framework of meaning by which I approach the world. The breadth of their provocation also included reconsiderations of the thoroughly concrete: why is that rubble, and that beauty; is their presence a sad victory in the fight of "man versus nature;" what is this dichotomy, because what, precisely, is a garden?

Gardens are synthetic; they "are natural places that have been transformed by creative human activity."[12] Yet, while their transformation makes them distinct

---

[8]   Siegfried Sassoon, *Siegfried's Journey 1916–1920* (New York: Faber & Faber, 1945).

[9]   Richard Reynolds, *On Guerrilla Gardening: A Handbook for Gardening without Boundaries* (London: Bloomsbury, 2008), p. 27.

[10]   David Tracey, *Guerrilla Gardening: A Manualfesto* (Gabriola Island, British Colombia, Canada: New Society Publishers, 2007), pp. 2–3.

[11]   Joe Robinson, "Guerrilla Gardeners Dig In," *Los Angeles Times*, May 29, 2008, http://articles.latimes.com/2008/may/29/home/hm-guerrilla29 (accessed July 13, 2010).

[12]   Cooper, *A Philosophy of Gardens*, p. 21.

from wilderness, their naturalness, in the natural attitude, is subsumed under their being a human product. In the everyday, we define a garden as a patch of land someone owns, purposefully cultivates and tends, with the purpose of showcasing flowers or producing edible matter. Gardening invokes a human subjectivity distinct from the natural world, one that rules over nature by ownership and deed, that is, with the intention to cultivate and produce. A garden, it seems, must have a creator with power triumphant over its matter. As Francis Bacon wrote, while "God Almightie first planted a garden," it is also "the purest of humane pleasures. It is the greatest refreshment to the spirits of man."[13] Gardening may permit humanity one more way that we may stand in His image: creators, ourselves, we rule over the created by ordering the landscape. Yet, Forrest Clingerman reminds us, "We are simultaneously transcending and situated beings, simultaneously active subjects without boundaries and passive objects within a definite place. It is not sufficient to understand nature as general … we must also account for our interaction with and transcendence over nature."[14] As McDuffie explains elsewhere in this volume (Chapter 11), a narrow view of humanity as dominating over nature hampers the current need for environmental awareness and immediate, intelligent recreation of our crumbling urban artifice into productive, healthy green.[15] The traditional definitions of gardens are cluttered by bias and must be rethought. We need to listen to Thoreau, when he wrote, "Gardening is civil and social, but it wants the vigor and freedom of the forest and the outlaw."[16] While the outlaw may seek to dominate as much as a creator, still, he knows that action only comes from his bending the strictures already rigidly set in the civil and natural playing fields.

The definition of a garden that prescribes one subject who creates the object that is the garden renders a further, unfortunate consequence in that the lack of ownership of land seems to relieve each of us from environmental obligation, even while it is clear that a healthy world and healthy communities demand our collective attention and responsibility. Gary Snyder, the beat poet and naturalist, captures the phenomenological collective interdependence of people and their environments:

---

[13]   Francis Bacon, "Of Gardens," in *The Effayes: Or Counsels Civill and Morall*, ed. Christopher Morley (Norwalk, CN: The Easton Press, 1980), pp. 148–56, p. 149.

[14]   Forrest Clingerman, "Beyond the Flowers and the Stones: 'Emplacement' and the Modeling of Nature," *Philosophy in the Contemporary World* 11, 2 (2004): p. 17.

[15]   Another concrete example of this dominance hindering environmental renewal was revealed by Los Angeles' political motion, the "Food and Flowers Freedom Act" (introduced by the City Council President Eric Garcetti, July 8, 2009), that sought to overturn a 1946 municipal code that prohibited "truck gardens," the growing of vegetables in residential zones for sale off-site, which exemplified the dictation of a limited definition of gardening that stands in sharp contrast to urban-environmental sustainability.

[16]   Henry David Thoreau, "A Week on the Concord and Merrimack Rivers (1849)," in *The Writings of Henry David Thoreau*, vol. 1 (New York: Houghton Mifflin, 1906), p. 55.

> We are all indigenous to this planet, this mosaic of wild gardens we are being
> called by nature and history to reinhabit in good spirit ... To restore the land one
> must live and work in a place. To work in a place is to work with others. People
> who work together in a place become a community, and community, in time,
> grows a culture. To work on behalf of the wild is to restore a culture.[17]

In the natural attitude, the daffodils on a street corner do not constitute a garden: who owns them? The flowers, thus freed from justification and definition, seem to be a poor example in the advocacy of guerrilla gardening as a phenomenologically-inspired environmental recreation. But, if we, without bias, openly approach the ownerless patch of daffodils, we can then hear how they call to us to step outside of our narrow egoism, how they call us to the phenomenological attitude, grant us a moment of beauty and ecological encouragement, and push us to further, positive recreation led by our wide-eyed, intentional relation in and with the world.

If we suspend our biases, set aside the privilege we give to a dominating encounter of our environment, and let the many modes of interpretation of each experience complement one another, reveal the multi-faceted dynamism of reality, then phenomenology can permit us a strategy for beautifying our surroundings in an environmentally responsible way (because, of course, the end goal is not a hundred randomly planted bulbs). To be able to see abandoned or ill-used land as potential events and sites that work for the good, however broadly these events, sites, and good ends may be defined, will be an affirmative step in the right and absolutely environmentally necessary direction. It is actually the numerous and radically diverse motivations behind guerrilla gardening, and the even greater number of its potential instantiations, that further necessitates the phenomenological study of guerrilla gardening while simultaneously making this illicit horticulture into a preeminent example of how phenomenology can assist urban renewal. Some city-dwellers plant abandoned lots to discourage crime and create community. For example, the garden activist Liz Christy and her band of Green Guerillas who founded, however unlawfully, the first community garden in New York City in 1973.[18] Green spaces can off-set greenhouse gases, control erosion, and replenish wildlife habitat. Some plant public spaces for a means of combating hunger. The city of Cleveland, Ohio, has consistently produced one million dollars worth of produce per year in its 200 plus community gardens.[19] And some work alongside

---

[17] Gary Snyder, "The Rediscovery of Turtle Island," in *A Place in Space: Ethics, Aesthetics, and Watersheds* (Washington, DC: Counterpoint, 1995), pp. 236–51, p. 250.

[18] Liz Christy Community Garden, www.lizchristygarden.org/ (accessed July 13, 2010) and Green Guerillas, www.greenguerillas.org/ (accessed July 13, 2010). Jacob's Well, a religious community group, similarly co-opted a vacant lot in Vancouver, British Colombia and funded its green re-development through a Community Supported Agriculture program (Tracey, *Guerrilla Gardening*, pp. 171–3).

[19] "Protecting Urban Gardens," EcoCleveland, 2002–03, www.ecocitycleveland.org/ smartgrowth/openspace/gardens.html (accessed July 13, 2010).

farmers to encourage organic, local production and support the remaining small farms. This public support of private farms can be seen in the dramatic rise in popularity of Community Supported Agriculture (CSA), where interested parties invest money or labor prior to harvest for shares in the expected yield.[20]

Even the smallest acts of horticultural insurrection, like the few daffodils on my local street corner, which will neither deter crime nor feed even a single hungry squirrel, do not lack profound significance for their ability to make us stop and think. Some guerrillas will scatter seeds along roadways for the encouragement of biodiversity, or simply, on a whim, to satisfy their own desires to beautify the long stretches of city concrete. Alone, such acts will not noticeably off-set greenhouse gasses, just as my daffodils will not protect the city corner from erosion. Many will simply plant ground they do not own because they can bring beauty and increase green space within a place that urban distraction has left littered with cigarette butts, needles, an occasional hubcap, and used condoms. The aesthetic reward may be enough for most guerrillas, but it is not the totality of meaning of even the smallest gardening act. "A community garden can be a school, church, nursery, playground and laboratory. It can also be a fitness center, picnic spot, habitat refuge, recycling depot and meeting place."[21] David Tracey does not mean that a garden can be *at* any of these places, but that it is, itself, an event that can teach, be spiritual, take care of, provide for, and more. An illicit garden, like the world at large, a random ecological act, like any action, does not present only a single meaning or reason for its origin, its being, or end. Instead, they give themselves to us as events with potential linkages to an entire network of meanings, if we are capable of thinking through their possible horizons. Within the natural attitude, guerrilla gardening is elusive to definition and can yield a product whose measurable environmental and social impact may quantitatively be near to nil. Nevertheless, the phenomenological impact of each guerrilla act, no matter how small, may be immense.

Consider the environmentally recreative potential that may come from even the smallest horticultural event: a driver sees the daffodils and smiles, becoming more courteous on the road, stomping less on the gas pedal. A child takes delight in them and encourages his mother to plant a small garden. A man is encouraged to get out and work in his yard, gaining good exercise, off-setting his risk of a heart attack. Families see the flowers, look more positively at the neighborhood, and consider moving back into the city center. Rambunctious college students see the flowers and are moved to buy bulbs to plant one midnight throughout the city, supporting their local garden centers. Commercially, a vendor sells more bulbs and a realtor sells more houses and the local economy improves. A girl wandering about spies them and is moved to reflect upon the meaning of human interaction

---

[20]   LocalHarvest, http://www.localharvest.org/csa/ (accessed July 13, 2010) and The United States Department of Agriculture, National Agriculture Library, http://www.nal.usda.gov/afsic/pubs/csa/csa.shtml (accessed July 13, 2010).

[21]   Tracey, *Guerrilla Gardening*, p. 138.

on and along with her environment and share such thoughts with others, who, too, may be induced to plant a small cluster.[22]

A chance encounter with a definition-challenging horticultural activity or site can provoke a phenomenological reflection upon how humanity and nature interact and influence one another. A subject, though, must approach the encounter with the proper attitude. Sometimes, one needs the shock of encountering the unexpected to initiate this proper mindset. And, if one's notions of gardens are too restrictive, one will be unable to openly encounter the random site of flowers or crops in a way that could positively effect meaning and affect the self. If one's regard for private property and laws governing the transformation of public spaces is too restrictive, one will look at the acts of guerrilla gardening as illegal or disobedient and not be open to their positive potential for generating societal change. Likewise, if one's regard is too permissive, then one may stimulate negative change by excusing non-responsible guerrilla acts (for example, the planting of hazardous or invasive species). Phenomenology, like most arts and all skills, needs to be practiced. This practice will not be divine mimicry, seeing ourselves as omnipotent as God, but will be more like a medieval spiritual exercise: loving the ideal that the event inspires, one repeatedly gives oneself to the world as the world gives itself to you so as to positively transform both world and self. Only when one steps outside of the biasing, everyday restraints of meaning and receptively approaches one's lived environment, can one then be moved, provoked, and encouraged to positive recreation by and to acts like those of guerrilla gardening.

Bibliography

Bacon, Francis, "Of Gardens," in *The Effayes: Or Counsels Civill and Morall*, ed. Christopher Morley (Norwalk, CN: The Easton Press, 1980), pp. 148–56.
Carr, David, "Husserl's Problematic Concept of the Life-World," in *Husserl: Expositions and Appraisals*, ed. Frederick Elliston and Peter McCormick (Notre Dame, IL: University of Notre Dame Press, 1977), pp. 202–12.

---

[22] Many of these benefits have been thoroughly studied and documented, for example: T. M. Waliczek, R. H. Mattson, and J. M. Zajicek, "Benefits of Community Gardening on Quality-of-Life Issues," *Journal of Environmental Horticulture* 14 (1980): pp. 204–9; Ishwarbhai C. Patel, "Gardening's Socioeconomic Impacts," *Journal of Extension* 24 (1991), www.joe.org/joe/1991winter/a1.php (accessed July 13, 2010); R. Marsh, "Building on Traditional Gardening to Improve Household Food Security," *Agriculture and Consumer Protection*, www.fao.org/docrep/X0051T/X0051T02.HTM (accessed July 13, 2010); D. J. Smith, "Horticultural Therapy: The Garden Benefits Everyone," *Journal of Psychosocial Nursing Mental Health Services* 36 (1998): pp. 14–21; and Rachel Kaplan, "Some Psychological Benefits of Gardening," *Environment and Behavior* 5 (1973): pp. 145–62.

Clingerman, Forrest, "Beyond the Flowers and the Stones: 'Emplacement' and the Modeling of Nature," *Philosophy in the Contemporary World* 11 (2004): pp. 15–22.

Cooper, David E., *A Philosophy of Gardens* (Oxford: Clarendon Press, 2006).

Heidegger, Martin, *Being and Time*, trans. Joan Stambaugh (New York: SUNY Press, 1996).

Husserl, Edmund, *Crisis of the European Sciences*, trans. David Carr (Evanston: Northwestern University Press, 1970).

Husserl, Edmund, *Ideas Pertaining to a Pure Phenomenology and to a Phenomenological Philosophy, First Book*, trans. F. Kersten (The Hague: Martinus Nijhoff Publishers, 1983).

Kaplan, Rachel, "Some Psychological Benefits of Gardening," *Environment and Behavior* 5 (1973): pp. 145–62.

Marsh, R., "Building on Traditional Gardening to Improve Household Food Security," *Agriculture and Consumer Protection*, www.fao.org/docrep/X0051T/X0051T02.HTM (accessed July 13, 2010).

Patel, Ishwarbhai C., "Gardening's Socioeconomic Impacts," *Journal of Extension* 24 (1991), www.joe.org/joe/1991winter/a1.php (accessed July 13, 2010).

Reynolds, Richard, *On Guerrilla Gardening: A Handbook for Gardening without Boundaries* (London: Bloomsbury, 2008).

Sassoon, Siegfried, *Siegfried's Journey 1916–1920* (New York: Faber & Faber, 1945).

Smith, D. J., "Horticultural Therapy: The Garden Benefits Everyone," *Journal of Psychosocial Nursing Mental Health Services* 36 (1998): pp. 14–21.

Snyder, Gary, *A Place in Space: Ethics, Aesthetics, and Watersheds* (Washington, DC: Counterpoint, 1995).

Thoreau, Henry David, "A Week on the Concord and Merrimack Rivers (1849)," in *The Writings of Henry David Thoreau*, vol. 1 (New York: Houghton Mifflin, 1906).

Tracey, David, *Guerrilla Gardening: A Manualfesto* (Gabriola Island, British Colombia, Canada: New Society Publishers, 2007).

Waliczek, T. M., R. H. Mattson, and J. M. Zajicek, "Benefits of Community Gardening on Quality-of-Life Issues," *Journal of Environmental Horticulture* 14 (1980): pp. 204–9.

Chapter 6

Eschatology of Environmental Bliss in Romans 8:18–22 and the Imperative of Present Environmental Sustainability from a Nigerian Perspective

Sampson M. Nwaomah

Introduction

Since the historian Lynn White's much-quoted article, "The Historical Roots of our Ecologic Crisis,"[1] in which he opined that the exploitive activities and negative ecological practices of technology derive from the biblical mandate "to rule and subdue the earth," there has been a great deal of discussion about the biblical mandate on the environment. White wrote that God's design for humanity was "to exploit nature in a mood of indifference to the feelings of natural objects."[2] He further argued, based on the account of Genesis 1:26–28 for humans to "rule and subdue the earth," that "it is God's will that man exploit nature for his proper ends."[3]

While it might easily be shown that the Old Testament (OT) passages on ecology (e.g. Gen 1:26–28; Ex 23:10, 11; Lev. 25:1–12; Ps 24:1–2; Ps 65:9) properly interpreted indicate responsible stewardship in the care of the environment,[4] some New Testament (NT) scholars like E. Lucas[5] argue that this may not be true of the NT. In his view, the difference between the NT's seeming

---

[1]   Lynn White Jr., "The Historical Roots of Our Ecologic Crisis," *Science* 155 (1967): pp. 1203–7.

[2]   *Ibid.*, p. 1205.

[3]   *Ibid.*

[4]   See Sampson M. Nwaomah, "Biblical Ecology of Stewardship: Options in the Quest of a Sustainable Environment in the Niger Delta Region of Nigeria," *Living Word: Journal of Philosophy and Theology* 113 (2007): pp. 89–103 and Sampson M. Nwaomah, "Water in the Bible in the Context of the Ecological Debate in the Nigerian Delta," *The Journal for Faith, Spirituality and Social Change* 1 (2008): pp. 187–204.

[5]   E. Lucas, "The New Testament Teaching on the Environment," in The John Ray Initiative (ed.), *A Christian Approach to the Environment* (Oswestry: John Ray Initiative, 2005), pp. 73–4.

indifference about environmental matters and the OT calls for stewardship might be because of the differences in the societies each of these collections of writings addressed. He holds that while the OT has a national and political community living on its own land as its audience, the Christian Church, the NT's audience, was a multinational community, with no political power and no specific land to lay claim to. Further, while the OT society had a strong agricultural base for its economy, which provides a background for awareness and an imperative of environmental sustainability, the NT society was largely urbanized and personal and interpersonal ethical matters dominated its writings. R. Bauckham, however, argues that the NT community was not as urbanized as Lucas contends. Further, rather than being environmentally detached, it might instead have been ignorant of the impact of "serious deforestation and desertification which some ancient farming was already causing, or of the extinction of species by human action."[6] Rather the dominant aspect of the environmental concern to the NT world was the threat of nature—"earthquake, shipwrecks, drought and famine, locust plagues, dangerous wild animals, animals that steal livestock or crops, extreme cold and extreme heat, and undrinkable water."[7]

Further, although it might be argued that some NT passages that refer to the environment do so informally and even so from an eschatological perspective (see Matt 5:18; Rom 8:18–22; 2Pt 3:7–13; Rev 21:1–2), it is, however, difficult to conclude that the divine care shown for non-human creation (Matt 6:25–30), the congeniality between Jesus and wild beasts (Mk 1:13), and the accountability that would be required of humankind for environmental management (Rev 11:18) do not point to an NT concern for environmental sustainability. An objective hermeneutics of these passages, and in fact every other NT passage on the environment, indicates that the six principles of eco-justice articulated by Norman Habel could apply to these passages. These principles are: (1) the principle of the intrinsic worth of all creation; (2) the principle of the interconnectedness of all creation; (3) the principle of the voice of creation; (4) the principle of the purpose of all creation; (5) the principle of the mutual custodianship by responsible custodians; and (6) the principle of the resistance by the earth against its abuse by human creation.[8]

Nevertheless, it might be thought that the NT's eschatological perspective on the temporality of the earth and its system of things encourages a full and unrestricted exploitation of the environment. For instance, if the present suffering of creation will give way to a new heaven and a new earth as implied in certain scriptural

[6]   R. Bauckham, "The New Testament Teaching on the Environment: A Response to Ernest Lucas," in The John Ray Initiative (ed.), *A Christian Approach to the Environment* (Oswestry: John Ray Initiative, 2005), pp. 97–9.

[7]   *Ibid.*

[8]   Norman C. Habel, "Introducing Ecological Hermeneutics," in Norman C. Habel and Peter Trudinger (eds), *Exploring Ecological Hermeneutics* (Atlanta: The Society of Biblical Literature, 2008), pp. 1–8.

passages such as Romans 8:18–22, why must Christians show concern about the present earth? Of what relevance is the principle of responsible stewardship in harnessing the resources of the environment? This chapter investigates Romans 8:18–22, in the context of the environmental quandary of the Niger Delta. Does the present suffering and decay or corruption of creation and the eschatology of environmental bliss in this passage justify current exploitative activities which disregard sustainability? Should this passage be interpreted in isolation from other biblical teachings on the environment? Does the scripture teach environmental domination in anticipation of the eschatology of environmental bliss regardless of the consequences for man? Are there no ethical responsibilities in economic pursuits? How should Christians in Nigeria relate to the ecological burden arising from the oil and gas industries? In my attempt to address these questions, this chapter is divided into five main sections: (1) an introduction; (2) a synopsis of the ecological situation in the Niger Delta; (3) a discussion of Romans 8:18–22 in the context of biblical teachings on the environment; (4) some recommendations for environmental sustainability in the Niger Delta Region of Nigeria; and (5) a conclusion.

A Synopsis of the Ecological Situation in the Niger Delta

Oil Exploration Activities in Nigeria

The quest to harness oil and gas resources in the Niger Delta dates back to 1909 when initial oil exploration began in the southern part of Nigeria—what is now know as the Niger Delta. Following the formation of the Nigerian state in 1914 and the Mineral Acts of the same year by which the central government declared autonomy over oil resources in the Nigerian territory, licenses and leases were granted to British companies and individuals to prospect for oil. Thus in 1937 the entire Niger Delta area was given to Shell to prospect for oil. Shell struck its first oil deposit in commercial quantity in Oloibiri (now Bayelsa State) in 1956 and later in Afam, Bomu and Ebubu (now in Rivers State). Oil exports began in 1958. Today, no fewer than 14 oil prospecting and exploitation companies with 159 oil fields and 1,481 producing wells are operational in Nigeria.[9] Oil revenues have historically provided about 95 percent of Nigeria's foreign exchange earnings and about 85 percent of federal revenue. This makes oil production central to the survival and effective functioning of the Nigerian state. A decline in oil production and/or fall in oil price on the international market both pose a significant threat to the nation's economy.

---

[9] B. O. Okaba, *Petroleum Industry and the Paradox of Rural Poverty in the Niger Delta* (Benin City: Ethiope Publishing, 2005), pp. 10–11.

Environmental Conditions in the Niger Delta

The 1998 Niger Delta Environmental Survey identified oil production and other industrial activities, population growth, agriculture, logging and fishing, as some of the factors that have had a significant impact on the development of the Niger Delta.[10] But ever since the discovery of oil in Olobiri (Bayelsa State) in 1956, the issue of oil production and its attendant consequences on the environment have produced continuous frictions between the oil companies and the host communities. Oil-based environmental problems are caused by oil spills, gas flaring, dredging of canals, barrow pits, and land acquisition for construction facilities. B. O. Okaba sums up the environmental impact of the oil industry in the Niger Delta of Nigeria as land deprivation, soil quality alteration, damage to the aquatic ecosystem, and air pollution.[11] This section of the chapter gives a précis of these effects.

Land deprivation   Land deprivation is occasioned by the physical impacts resulting from land preparations for seismic drilling. These activities include, but are not limited to, construction of flow lines and trunk line networks, terminals, digging of location waste pits and barrow pits. These activities require huge land areas and as a result most communities have been deprived of arable land for agriculture, the minimal compensations notwithstanding.

Soil quality alteration   Soil quality alteration is another environmental problem that the Niger Deltans contend with as a result of the oil industry. The direct cause of this is the oil spillage either from equipment failure, human error, or corrosion of pipes due to age and sabotage. It is estimated that as of 2002 about 8,581 oil spills, involving nearly 28 million barrels of oil, have occurred in the Niger Delta.[12] The last couple of years have witnessed several hundred more oil spills and some certainly have gone unreported. Other factors responsible for the soil quality alterations are the unceasing gas flares, drill cuttings, drill mud and refinery waste. The harmful effects of these elements on the soil are unquestionable. Several studies have demonstrated this. S. W. Owabukeruyele argued that the compounds from the numerous petroleum wastes contain organic chemicals such as phenol cyanide, sulpheide-suspended solids, chromium and biological oxygen that have

---

[10]   C. Ile and C. Arukwe, ""Niger Delta, Nigeria: Issues, Challenges and Opportunities for Equitable Development," retrieved from: http://nigeriaworld.com/feature/article/niger-delta.html on February 1, 2009, par. 10.

[11]   Okaba, *Petroleum Industry*, pp. 18–20.

[12]   *Ibid.*, p. 15.

destructive effects on the land and water.[13] Owabukeruyele's claim is corroborated by other research.[14]

Damage to aquatic ecosystem   Another environmental tragedy in the Niger Delta is the damage to the aquatic ecosystem. This aspect is quite significant to the people because it constitutes a grave threat to the traditional economic hub of their lives—fishing. It is a common practice that in the process of oil exploration and production materials such as drill cuttings, drill mud and other fluids that are used to stimulate production are discharged into the environment and most end up in the streams and rivers. These chemicals are not easily degradable. Skimming off oil from the water surface hardly solves the problem, since most of the oil might have sunk to the bottom of the water, producing such destructive environmental consequences as: (1) surface and ground water quality deterioration in terms of drinkability, aesthetics, and recreation; (2) destruction and reduction of fish and fisheries production in the waters; (3) destruction by acute and sublethal toxicity of aquatic flora and fauna, including benthic macroinvertebrates, because of the toxins discharged into the water.[15]

Air pollution   Lastly, there is the hazard of air pollution. The obvious cause of this is the gas flares from oil production in numerous flow stations in the Niger Delta. The pollutants released and the noise resulting from the process both are injurious to human health. The heat emitted from the flares affects the adjacent farmlands, scares wildlife and contributes to the pollution of nearby streams, which sometimes are the only source of drinkable water for local communities. The chemical emissions from flaring contribute to acid rain and might be a cause of skin diseases.[16] The emissions also cause the quick corrosion of roofing sheets on houses in the Niger Delta area, a common sight in this region. A secondary source of air pollution in the Niger Delta is the incidents of fire resulting from leakages from the exposed and corroded oil pipes that are scattered all over the region.

---

[13]   S. W. Owabukeruyele, "Hydrocarbon Exploitation, Environmental Degradation and Poverty in the Niger Delta of Nigeria," paper presented at the Lund University LUMES Program, Lund, Sweden in 2000. Retrieved from: www.waado.org/Environment/PetrolPolution/EnvEconomics.htm on February 1, 2009, par. 3.

[14]   See A. O Isiche, and W. W. Stanford, "The Effect of Waste Gas Flares on Surrounding Vegetation of South-eastern Nigeria," *Journal of Applied Ecology* 13 (1976): pp. 177–87; N. Nwankwo, and C. N. Ifeadi, "Case Studies on the Environmental Impact of Oil Production and Marketing in Nigeria" (University of Lagos, Nigeria, 1988); G. O. Sanni and S. O. Ajisebutu, "Effects of Chronic Exposure of Soil and Soil Bacteria to Petroleum Fractions," *Environtropica: An International Journal of the Tropical Environment* 1 (2004): pp. 119–28.

[15]   Okaba, *Petroleum Industry*, pp. 19–20.

[16]   *Ibid.*, pp. 21–2.

Romans 8:18–22 in the Context of Biblical Teachings on the Environment

We begin our discussion of Romans 8:18–22 with the passage:

> 18 For I consider that the sufferings of this present time are not worthy to be compared with the glory which shall be revealed in us. 19 For the earnest expectation of the creation eagerly waits for the revealing of the sons of God. 20 For the creation was subjected to futility, not willingly, but because of Him who subjected it in hope; 21 because the creation itself also will be delivered from the bondage of corruption into the glorious liberty of the children of God. 22 For we know that the whole creation groans and labors with birth pangs together until now. (NKJV)

Earlier in Romans 8, Paul discusses the sufferings that are present in the world. In this passage, Paul attempts to show that in these sufferings, three sources of encouragement are available to Christians: (1) a future hope (vv 18–25); (2) a present help (vv 26, 27); and (3) the past election of God (vv 28–30). Commenting on this section (Rom 8:18–39) to which our passage belongs, Ben Witherington III and Darlene Hyatt opine: "It reveals one of the most masterful dimensions of Paul's theology. Paul shows here that salvation in Christ completes not only God's plan for creation, but also his plans for calling and forming people for himself. Thus for Paul salvation has both anthropological and cosmological significance and effects."[17]

In the same vein, Richard C. Halverson specifically notes that Romans 8:18–25 "is an exciting glimpse into Paul's philosophy of history; one of his definite insights into the Biblical view of history."[18] Similarly, Brendan Byrne (as cited by Sigve Tonstad) holds that this Pauline passage appears to be his distinctive account of "human beings in relation to the nonhuman created world."[19] Thus, this passage is considered a commentary on Paul's theology of eschatological bliss for both human and nonhuman creation. For our purpose, discussion will center on four key words in this passage. These words are: (1) creation, (2) corruption, (3) longing, and (4) restoration. No elaborate exegetical process is adopted here.

---

[17] Ben Witherington III and Darlene Hyatt, *Paul's Letter to the Romans: a Socio-Rhetorical Commentary* (Grand Rapids: William B. Eerdmans, 2004), p. 221.

[18] Richard C. Halverson, *The Gospel for the Whole Life: The Book of Romans in Today's World* (Palm Springs: Ronald N. Haynes Publishers, 1981), p. 65.

[19] Sigve Tonstad, "Creation Groaning In Labor Pains," in Norman C. Habel and Peter Trudinger (eds), *Exploring Ecological Hermeneutics* (Atlanta: The Society of Biblical Literature, 2008), p. 142.

Creation and Corruption

The meaning of the word *ktisis* (translated in the NKJV as "creation" in verses 19–22) has elicited major concerns. Witherington and Hyatt identify eight possibilities namely: (1) all humanity, (2) unbelieving humanity alone, (3) believing humanity alone, (4) angels alone, (5) nonhuman nature (both creature and creation), (6) subhuman nature plus angels, (7) unbelievers and nature, and (8) subhuman nature plus humanity in general.[20] However, the most plausible meaning of "creation" in this passage points to nonhuman creatures and nature which were subjected to the futility, corruption, and pain, experiences described in the passive voice. This experience is possibly linked to Genesis 3:17–18, the creation story of the fall of Adam and Eve and its effect on nonhuman creatures and creation. Paul seems to imply here a working of the law of thermodynamics, in particular the movement toward entropy. The passage reads,

> Then to Adam He said, "Because you have heeded the voice of your wife, and have eaten from the tree of which I commanded you, saying, 'You shall not eat of it': *Cursed is the ground for your sake*; in toil you shall eat of it all the days of your life. 18 Both thorns and thistles it shall bring forth for you, and you shall eat the herb of the field." (NKJV, emphasis mine)

Consequently, the sin of Adam opened the way for the operation of this law such that nature tended toward decay and death.[21] The futility (an inability or failure to achieve intended aim) and decay or corruption (an idea of weakness and death as opposed to glory) of creation limits creation's power to achieve the purpose God intends for it. The thorns and thistles, symbols of futility and decay, as Tonstad opines, "may even be seen as the outward expression of inward resistance"[22] by creation because of the effect of the Fall on it. Expounding on this theme, John Bowen suggests that "lands which were fertile in the past, and offered the prospect of good harvest for many years, have become desert. Sometimes this resulted from the change of climate, sometimes from people's use of destructive weapons, sometimes from their selfishly or ignorantly over-grazing the land with too many cattle, or cutting down too many trees,"[23] and of course other human activities like global warming and the disregard of environmental best practices in economic and technological pursuits. In this vein the *Fall's expression of curse, decay, ineffectuality and restraint* on material creation in Genesis 3:17, 18 reverses the *expression of beauty, fruitfulness abundance, and harmony* in Genesis 2:7–15 as described below:

---

[20]   Witherington III and Hyatt, *Paul's Letter to the Romans*, pp. 222–3.

[21]   John L. Benson, *Romans: Life of the Justified* (Denver: Accent Publications, 1972), p. 59.

[22]   Tonstad, "Creation Groaning In Labor Pains," p. 143.

[23]   Roger Bowen, *A Guide to Romans* (London: Redwood Books, 1996), p. 122.

7 And the LORD God formed man of the dust of the ground, and breathed into his nostrils the breath of life; and man became a living being. 8 The LORD God planted a garden eastward in Eden, and there He put the man whom He had formed. 9 And out of the ground the LORD God made every tree grow that is pleasant to the sight and good for food. The tree of life was also in the midst of the garden, and the tree of the knowledge of good and evil. 10 Now a river went out of Eden to water the garden, and from there it parted and became four riverheads. 11 The name of the first is Pishon; it is the one which skirts the whole land of Havilah, where there is gold. 12 And the gold of that land is good. Bdellium and the onyx stone are there. 13 The name of the second river is Gihon; it is the one which goes around the whole land of Cush. 14 The name of the third river is Hiddekel; it is the one which goes toward the east of Assyria. The fourth river is the Euphrates. 15 Then the LORD God took the man and put him in the garden of Eden to tend and keep it. (NKJV)

Longing and Restoration

However, the present sufferings and bondage of nature notwithstanding, Paul—in typical Jewish fashion—looks forward to the restoration of material creation to its original nature before the Fall and its attendant consequences (Rom 8:20b–22). This "eager expectation" (from the Greek *apokaradokia*, a word used in describing one leaning forward out of consuming interest and desire for a realization of hope) may be built on the Jewish theology of restoration as illustrated in Baruch, for example: "The vine shall yield its fruit ten thousand fold, and on each vine there shall be a thousand branches; and each branch shall produce a thousand clusters; and each cluster produce a thousand grapes; and each grape a cor of wine. And those who have hungered shall rejoice; moreover, also, they shall behold marvels every day" (2 Baruch 29:5, 6).

This concept of future environmental bliss is also evident in the great vision of Isaiah 11:1–10. In this vision, no creature kills for sustenance and there is no war or injustice in human society. This reconciliation between humanity and the rest of creation evokes a return to the Garden of Eden. W. Shedd, as cited by Harrison, observes: "the restoration of material nature is a condition similar, in its own lower sphere, to the restoration of man's spiritual nature, in its higher sphere. St Paul here teaches, not the annihilation of this visible world, but its transformation."[24]

Arising from this Pauline—and more generally New Testament (cf. Matt 5:18; Rom 8:18–21; 2Pt 3:7–13; Rev 21:1–2)—eschatological optimism for the material or nonhuman creation is the purposefulness of nature. Commenting on this, Tonstad writes: "Paul's New Testament witness reveals that there is a purpose

---

[24]   Everett F. Harrison, "Romans," in Frank E. Gaebelein (ed.), *The Expositor's Bible Commentary*, Vol. 10 (Grand Rapids: Zondervan Publishing House, 1976), p. 94.

for nature, too: a God-ordained bill of rights. Even if the purpose is temporarily thwarted by human sin, nature is not left without hope."[25]

However, influenced by White, some assume that this eschatological hope of nature gives humankind the liberty to explore and "exploit nature in a mood of indifference to the feelings of natural objects," as pointed out above. Thus the abuse and destruction of the environment in the quest for its natural reserves seems biblically justified. It is like saying, "use this up; for the new shall come."

While the scriptural allusions to the eager expectation of a blissful environment may appear to give latitude to those who seek to abuse the environment in the quest for its riches, I seek to show that this text cannot be interpreted in isolation from other biblical teachings on God's pre- and post-Fall relationship to, and/or stance on, the environment. In discussing this we should investigate (1) the status of created nature and the restrictions placed on its usage and (2) the relationship of human uniqueness to the environment.

The Status of Non-Human Creation and Restrictions on its Usage

The language of the creation account in Genesis 1:31 ascribes "goodness" to all the creatures of God. This goodness extends to nature in as much as it also reflects the work of the creator. Further, God is depicted as caring for all creation—human and nonhuman. He waters "a land where no man lives, a desert with no one in it, to satisfy a desolate wasteland" (Job 39:26–27), and provides enabling environment for wild animals (Job 39:5–6; Ps 104:10–27). In the New Testament God cares for the birds and the lilies (Matt 6:26–30, 10:29; Lk 12:6). Jesus, in his wilderness experience, was reported to be at home with wild animals (Mk 1:13). God's care for this level of creation is predicated on the consistent theme of God's creatorship and ownership of the earth and all that dwell in it (Ps 24:1, 50:10, 89:11; Exod. 19:5; Deut. 10:14; Job 38–41). Clearly such depictions of nonhuman creation's status imply that their value is beyond the utility they provide for humans.

In this context, it may be relevant also to recall some of the restrictions given on the usage of nonhuman creation. For instance, the so-called "dominion" mandate given to humankind described in Genesis 1 permits humans to use only seed-yielding plants and trees for food, while animals are given "every green plant" to eat (Gen. 1:29–30). The fruit of trees is not to be eaten until the fifth year after planting; all fruit from the fourth year is to be "set apart for rejoicing in the Lord" (Lev. 19:23–25). Fruit trees are also to be spared during time of war, even when their wood is needed for the construction of siege works (Deut. 20:19–20). Fields and vineyards are to be left fallow every seventh (sabbatical) year and fiftieth (jubilee) year—"a year of complete rest for the land" (Lev. 25:1–12). And tithes of the land's produce belong to the Lord (Lev. 27:30). In sum, the rather circumscribed uses of natural resources authorized to human beings in the Old Testament would

---

[25]   Tonstad, "Creation Groaning In Labor Pains," p. 143.

hardly support any notion of human "dominion" as a license to exploit.[26] There is an eschatological echo that God disavows an abuse of the environment (Rev 11:18). Clearly, then, the status of nonhuman creation, God's concern for them both in the pre- and post-Fall narration, and the restrictions placed on their usage do not imply that their function is intrinsically anthropocentric.

Human Uniqueness in Relation to Nonhuman Creation

In addition to human usage of nonhuman creation while awaiting the eschatology of bliss, a second point worth discussing is the "dominion" mandate given in Genesis. In Genesis 1:28–29, humans are told to "be fruitful and multiply, and fill the earth and subdue it," to have "dominion" (*radah*) over "every living thing that moves upon the earth," and to eat from every seed-bearing plant and tree. This language of supremacy of humans in the relationship to the earth seems buoyed by the belief that Adam and Eve were described as created "in the image of God," a characteristic or attribute not attributed to the other nonhuman creatures. But what does this status mean in terms of humankind's relationship to the rest of the natural world? Can it be taken to either imply authorship or ownership of nature? Does it warrant usage of natural resources without restriction?

The nature and extent of this mandate is contentious. In one sense, in Genesis the categorization of creation moves successively from its lowest to its highest forms, the last (the human) to be served by all the former as implied in the passage, and also by the Yahwist creation account (Gen. 2:4b–25). Thus the creation of plants, animals and the natural resources are *subordinate to* Adam's creation, that is, the rest of creation is made in order to serve humankind. Hence the existence of the nonhuman creation is *anthropocentric* rather than *theocentric*. This perception of the nonhuman creation led White to argue: "God planned all of this explicitly for man's benefit and rule: no item in the physical creation had any purpose save to serve man's purposes."[27] Following this line of interpretation of the value of nonhuman creation, Thomas Aquinas argued that only the intellectual creature (the human) has "the essential character of a principal agent," among created beings, while non-intellectual creatures have "the formal character of an instrument" and, thus, are not valued for their own sakes but "as useful to a principal agent."[28] He goes on to point out that nonhuman creation thereby stands in instrumental subjection to the human. Earlier Origen had opined that the creator "has made everything to serve the rational being and his natural intelligence," and similarly the Stoic who posited the creator, "quite rightly put man and the rational nature in

---

[26]   James B. Tubbs, Jr., "Humble Dominion," retrieved from: http://theologytoday. ptsem.edu/jan1994/v50-4-article4.htm on January 16, 2009, p. 545.

[27]   White Jr., "The Historical Roots of Our Ecologic Crisis," p. 1205.

[28]   Aquinas cited in Tubbs, Jr., "Humble Dominion," p. 545.

general above all irrational beings," and says that providence has made everything primarily for the sake of the rational creature.[29]

In contrast to this position, it has also been argued that "dominion" is a fraternal term in the Old Testament. Commenting on the concept to "subdue and have dominion over," Hamilton maintained that in the 24 usages in the Old Testament of "dominion," it is "to be exercised with care and responsibility. Nothing destructive or exploitative is permissible."[30] Commenting also on this passage, G. J. Wenham writes that the creation of humanity in God's image, and thus as his representative on earth, requires humanity "to act in a godlike way in caring for the earth and other creatures in it."[31]

It seems that the "image of God" denotes the desire of God for humans to reflect certain divine capabilities and the fulfillment of his purposes for nonhuman creation in human purposes. Humans are therefore in "God's image" really only as God's representative, called to preserve and enforce God's claim to dominion over the earth.[32] In this sense, therefore, the environment is a gift given to humans for sustenance, and the key elements in this human/nonhuman relationship are stewardship, responsibility, and sustainability of the environment that has been entrusted to humanity. Destructive, manipulative, indiscriminate, and exploitative activities and policies damaging to host communities (all of which are seen in the case of the Niger Delta) are diametrically opposed to the stewardship and dominion concept of the biblical ecology.[33]

Recommendations for Environmental Sustainability in the Niger Delta Region of Nigeria

In the light of the insights gained from the above discussion of the status of nonhuman creation and the restrictions God placed on humans in relation to their usage, his care for them, and the stewardship mandate, rather than exploitation being permitted to man when God gifted him with the environment, it is my position that it is a limitation of theological insight to perceive Romans 8:18–22, and other NT eschatological references to the environment, as an authorization to recklessly use the environment in anticipation of an eschatology of bliss. Rather than Paul sanctioning man's hostility toward nonhuman creation and vice versa, he argues that nature suffers because of humanity and humanity may, through best practices, in the interim liberate it.

---

[29] *Ibid.*

[30] V. P. Hamilton, *Handbook on the Pentateuch* (Grand Rapids: Baker Book House, 1982), pp. 27–8.

[31] G. J. Wenham, *Exploring the Old Testament: A Guide to the Pentateuch* (Downers Grove: InterVarsity Press, 2003), p. 20.

[32] Gerhard von Rad, *Genesis* (Philadelphia: Westminster, 1972), p. 60.

[33] Nwaomah, "Biblical Ecology," p. 92.

Following this, therefore, I seek to make recommendations toward environmental sustainability in the Niger Delta Region of Nigeria—a region that is plagued with a high level of infrastructural deficit, unemployment, poverty, illiteracy, ignorance, disease, and alarming evidence of all the indices of under-development resulting in violence, crime, and cultism and recently kidnapping.[34] In these recommendations, it is my position that the Christian community in Nigeria should be in the vanguard in campaigning for sustainable environmental practices. This is so because Christians, according to the General Board of Church & Society of the United Methodist Church, are to "recognize the responsibility of the church and its members to place a high priority on changes in economic, political, social, and technological lifestyles to support a more ecologically equitable and sustainable world leading to a higher quality of life for all of God's creation."[35]

Diversification of the Economy of Nigeria

As pointed out earlier in this chapter, oil revenues have historically provided about 95 percent of Nigeria's foreign exchange earnings and about 85 percent of federal revenue. This makes oil production central to the survival and effective functioning of the Nigerian state. However, it is common knowledge that oil and gas are exhaustible natural resources and extreme dependence on them for national wellbeing could jeopardize the future of a nation which has neglected to diversify its economy. Nigeria is blessed with huge natural resources such as tin, iron ore, coal, limestone, niobium, lead, zinc, wildlife, and arable land for agriculture, which can significantly complement the oil and gas industry and conserve the environment and its resources for future generations. As I have argued elsewhere: "A well-managed environment secures the sustenance of posterity … The diversification of economic opportunities will curtail the sustained seismic activities in search of oil and gas deposits in the Niger Delta and secure the interests of the posterity of the Region and its people."[36]

---

[34]   Chibuike Rotimi Amaechi, "Phases in the Manifestation and Management of the Niger Delta Crisis," paper presented at the University of Ilorin, second distinguished personality lecture organized by the Centre for Peace and Strategic Studies (CPSS), of the University of Ilorin, Kwara State, March 23, 2009.

[35]   General Board of Church & Society of the United Methodist Church "Economic and Environmental Justice," retrieved from: www.umcgbcs.org/site/c.frLJK2PKLqF/b.2808983/k.447D/Economic_and_Environmental_Justice.htm on February 1, 2009, par. 4.

[36]   Nwaomah, "Biblical Ecology," p. 100.

Formulation and Implementation of Environmental-Friendly Policies on Oil Exploration and Exploitation

Nigeria has an impressive array of environmental laws such as the Environmental Impact Assessment decree No. 86, 1992 and a Ministry of Environment, but little concern is given to the implementations of inconvenient environmental laws. For instance, the 1969 Petroleum Act (CAP 350), the first and the main oil-related statute, provides for the exploitation and exploration of petroleum in Nigeria and its territorial waters. Certain provisions in CAP 350 prohibited oil-based activities in certain areas. Specifically, Section 17 of the Act prohibits oil-based activities in the following areas:

1. any area held to be sacred;
2. any part set apart for or used or appropriated or dedicated to public purposes;
3. any part occupied for any purpose by government at the federal or state levels;
4. any part situated within a township, town, village, market, burial ground or cemetery;
5. any part which is the site of within 50 yards of any building, institution, reservoir, dam, public road or train way or which is appropriated for or situated within 50 yards of any railway;
6. any part consisting of private land;
7. any part under cultivation.

The implementation of the provisions as listed above is rarely respected by the oil companies in the Niger Delta. For instance, there are many sacred sites and/or areas for the people in the Delta region. But the absolute authority to determine what constitutes sacredness and which area(s) are sacred lies with the government and not the people. Oil is exploited and explored in residential areas and farmlands under cultivation are not spared. The determining index is the economic gain to the government and the prospecting and exploring companies. The provisions of Sections 25 and 36 of the Act, which touch on safety standards in petroleum drilling and production, are also hardly observed. These sections specifically focus on the prevention and control of pollution. Section 35 requires the licensees to ensure that minimal damage is caused to the surface of the relevant area, vegetation, structures and other property thereon. But no penalty is provided in case of violation and consequently the multinational oil companies brashly ignore these laws. It is estimated that about "2.300m³ of oil is spilled in 300 separate incidents annually in the Niger Delta."[37] Since by law oil and gas companies are obliged to pay compensation for spills from "deliberate, destructive acts," most companies are quick to point to sabotage as the cause for every spillage reported. Gas flaring

---

[37]   *Ibid.*, p. 120.

in Nigeria is unequalled. It is estimated that about 76 percent of Nigeria's total gas production is flared.

According to the Nigeria's Department of Petroleum Resource (DPR), Nigeria flared a total volume of 1.7 billion standard cubic meters of associated gas as against its total domestic and industrial use of 916 million standard cubic meters. The Nigerian government has not been able to enforce the legislation against gas flaring and its conversion to domestic and profitable industrial use. But for the health of the people, it is needful for all the relevant agencies, such the Environmental Protection Agency of the Ministry of Environment, to enforce policies that can positively impact on the Niger Delta environment and its people.

Environmental Justice

A third recommendation for the purpose of a present sustainable environment and anticipation of its eschatological redemption can be made from the perspective of environmental justice. In the words of Robert D. Bullard, "Environmental justice is the fair treatment and meaningful involvement of all people regardless of race, color, national origin, or income with respect to the development, implementation, and enforcement of environmental laws, regulations, and policies."[38] Fair treatment means that no group of people, including racial, ethnic, or socio-economic groups, should bear a disproportionate share of the negative environmental consequences resulting from industrial, municipal, and commercial operations or the execution of federal, state, local, and tribal programs and policies. This requires that all human beings, created in the image of God, are entitled to equal protection and equal enforcement of our environmental, land use and energy, laws and regulations. This leads to social responsibility and ethical considerations in the use of the environment.

These considerations become imperative in environmental sustainability when one considers the imbalance in technological advancement, policy-making and implementation, which not only impact on the present vulnerable generation but jeopardize the future generations. Consequently, any conservation ethic for the Niger Delta must consider the welfare and material needs of future persons. This ethical consideration is in harmony with nature/environment being a gift to all humanity in every generation until the liberation anticipated in Romans 8:18–22 is achieved.

---

[38]   Robert D. Bullard, "Environmental Justice for All," retrieved from: http://nationalhumanitiescenter.org/tserve/nattrans/ntuseland/essays/envjust.htm on January 16, 2009, par. 3.

Conclusion

In conclusion, the eschatology of environmental bliss which Paul writes about in Romans 8, if studied in the context of the biblical teachings on the environment both pre- and post-Fall, provides no warrant to recklessly explore and exploit the environment in the pursuit of economic gains or technological advancement. The perceived sovereignty of the human race over nonhuman creation is subordinated to the overall sovereignty of God over man and the mandate to care for God's creation. Further, the thought that human redemption is correspondingly linked with that of the nonhuman creation should awaken in us the imperative to engage in activities and policies that would enhance the sustainability of the environment in anticipation of eschatological redemption.

Bibliography

Amaechi, Chibuike Rotimi, "Phases in the Manifestation and Management of the Niger Delta Crisis," paper presented at the University of Ilorin, second distinguished personality lecture organized by the Centre for Peace and Strategic Studies (CPSS), of the University of Ilorin, Kwara State, March 23, 2009.

Benson, John L, *Romans: Life of the Justified* (Denver: Accent Publications, 1972).

Bowen, Roger, *A Guide to Romans* (London: Redwood Books, 1996).

Bullard, Robert D. "Environmental Justice for All," retrieved from: http://nationalhumanitiescenter.org/tserve/nattrans/ntuseland/essays/envjust.htm on January 16, 2009.

Habel, Norman C. and Peter Trudinger (eds), *Exploring Ecological Hermeneutics* (Atlanta: Society of Biblical Literature, 2008).

Halverson, Richard C., *The Gospel for the Whole Life: The Book of Romans in Today's World* (Palm Springs: Ronald N. Haynes Publishers, 1981).

Hamilton, V. P., *Handbook on the Pentateuch* (Grand Rapids: Baker Book House, 1982).

Gaebelein, Frank E. (ed.) *The Expositor's Bible Commentary*, 12 vols (Grand Rapids: Zondervan Publishing House, 1976).

General Board of Church & Society of the United Methodist Church "Economic and Environmental Justice," retrieved from: www.umcgbcs.org/site/c.frLJK2PKLqF/b.2808983/k.447D/Economic_and_Environmental_Justice.htm, on February 1, 2009.

Ile, C. and C. Arukwe, "Niger Delta, Nigeria: Issues, Challenges and Opportunities for Equitable Development," retrieved from: http://nigeriaworld.com/feature/article/niger-delta.html on February 1, 2009.

Isiche, A. O and, W. W. Stanford, "The Effect of Wastes Gas Flares on Surrounding Vegetation of South-eastern Nigeria," *Journal of Applied Ecology* 13 (1976): pp. 177–87.

The John Ray Initiative (ed.), *A Christian Approach to the Environment* (Oswestry: John Ray Initiative, 2005).

Nwankwo, N. and C. N. Ifeadi, "Case Studies on the Environmental Impact of Oil Production and Marketing in Nigeria" (University of Lagos, Nigeria, 1988).

Nwaomah, Sampson M., "Biblical Ecology of Stewardship: Options in the Quest of a Sustainable Environment in the Niger Delta Region of Nigeria," *Living Word: Journal of Philosophy and Theology* 113 (2007): pp. 89–103.

Nwaomah, Sampson M., "Water in the Bible in the Context of the Ecological Debate in the Nigerian Delta," *The Journal for Faith, Spirituality and Social Change* 1 (2008): pp. 187–204.

Okaba, B. O., *Petroleum Industry and the Paradox of Rural Poverty in the Niger Delta* (Benin City: Ethiope Publishing, 2005).

Owabukeruyele , S. W, "Hydrocarbon Exploitation, Environmental Degradation and Poverty in the Niger Delta of Nigeria," paper presented the Lund University LUMES Program, Lund, Sweden in 2000. Retrieved from: www.waado.org/ Environment/PetrolPolution/EnvEconomics.htm on February 1, 2009.

Sanni, G. O. and S. O. Ajisebutu, "Effects of Chronic Exposure of Soil and Soil Bacteria to Petroleum Fractions," *Environtropica: An International Journal of the Tropical Environment* 1 (2004): pp. 119–28.

Tonstad, Sigve, "Creation Groaning In Labor Pains," in Norman C. Habel and Peter Trudinger (eds), *Exploring Ecological Hermeneutics* (Atlanta: The Society of Biblical Literature, 2008).

Tubbs, Jr., James B. "Humble Dominion," retrieved from: http://theologytoday. ptsem.edu/jan1994/v50-4-article4.htm on January 16, 2009.

von Rad, Gerhard, *Genesis* (Philadelphia: Westminster, 1972).

Wenham, G. J., *Exploring the Old Testament: A Guide to the Pentateuch* (Downers Grove: InterVarsity Press, 2003).

White, Jr., Lynn. (1967) "The Historical Roots of Our Ecologic Crisis," *Science* 155 (1967): pp. 1203–7.

Witherington III, Ben and Darlene Hyatt, *Paul's Letter to the Romans: A Socio-Rhetorical Commentar* (Grand Rapids: William B. Eerdmans Publishing Company, 2004).

Chapter 7

Resurrecting Spirit: Dresden's Frauenkirche and the Bamiyan Buddhas

James Janowski

Introduction

On February 13–14, 1945, with World War II very nearly at its end, Allied bombs obliterated 15 square kilometers of Dresden, a city then known as "Florence on the Elbe" for its magnificent architecture, culture, and history. The Frauenkirche, an iconic jewel on the city's beautiful Baroque horizon, initially survived the pounding attack and subsequent firestorm. But the heat proved too much— temperatures inside the church reached 1,000 centigrade—and the building's infrastructure buckled and gave way. On February 15 at 10:00 a.m., two days after the bombs fell, the Frauenkirche collapsed into a pile of rubble. By the time the smoke cleared and the embers cooled, the choir section and a staircase, both striking a mournful pose, were the only two parts left standing.

Almost immediately, Dresdeners were drawn to the site and drawn, irresistibly, to safeguarding the remains of their church. Plans for its resurrection, too, began almost immediately. But these plans languished as the city, part of the German Democratic Republic (GDR), fell into economic and religious torpor. Thus the ruins remained in place, largely undisturbed, for nearly half a century. Indeed, it took the fall of the Berlin Wall—and the increased "openness" as well as access to financial resources that flowed from this—before the resurrection project, which had never languished in the minds of some Dresdeners, could commence. Fundraising began with the "Call from Dresden," a citizens' initiative and clarion call, on the 45th anniversary of the bombing. The call was hugely successful, with support coming in from all over the world, including the Western powers that had destroyed the church and its environs. The project began in 1993 and was completed in 2005. Twelve years in the re-making—and only five years less than the church had initially taken to construct—the resurrected Frauenkirche was re-consecrated on October 30, 2005, in time for the 800-year anniversary of Dresden's founding in 2006.

My aim is two-fold. First, I will describe and evaluate the reconstruction of the Frauenkirche, focusing on the religious and spiritual dimensions of the project.[1] (Obviously the project has other important dimensions—aesthetic, historical, political, and the like. I will allude to some of these, but space constraints preclude me from considering them with care.) Second, I will extrapolate from Dresden's case to another, more recent example of cultural barbarism—the Taliban's desecration, in 2001, of Afghanistan's Bamiyan Buddhas. The Frauenkirche and the Buddhas are powerful analogues; and I will seek to draw a moral of the Dresden story for Bamiyan.

The Frauenkirche: Its History, Significance, and Resurrection

Dresden's Frauenkirche can be traced back to the eleventh century—well before the founding of the city itself in 1206. Some version of the church, then, has been on this location for a millennium. The 1945 incarnation—designed by Dresden master carpenter George Bähr, financed by local donations (and beer taxes), and constructed by local laborers just over 200 years earlier—was a splendid and much-revered icon. The Frauenkirche was central to the history of Lutheranism. Indeed, Bähr's church—with its altar and pulpit center-stage, in full view of the entire congregation—powerfully portrayed Protestantism's egalitarian ideals. Germany's largest Protestant church was also rich in musical lore—Bach and Wagner performed there—and a witness to much history. (The church survived direct, repeated artillery strikes in the Seven Years' War and was used as a refuge in the May Uprising of 1849.) And then there is its architectural significance. The building's original construction unlocks part of the history of masonry.[2] Bähr's inspired plans, never mathematically precise or "proven," appealed to "Spieramen," a slender structural feature he believed would support the 220 foot high, 12,000 ton stone dome. (Some of Bähr's blueprint was "intuitive." Dresdeners' collective memory has it that the carpenter drew lines in the dirt, improvising on site, in overseeing the church's construction.) But Bähr's unique concept—this was the first and only time "Spieramen" has been used in architectural design and engineering—was insightful, ingenious, and prescient. He—and thus his building—anticipated technical developments in architecture that were to follow. In short, the Frauenkirche was magnificent and quirky. Those who restored it in

---

[1]   Using "reconstruction" here I follow most who discuss this case. Obviously "reconstruction" overlaps, in terms of meaning, each of "recreation," "restoration," and "replacement"—the focal terms of this volume. While it is not my central purpose here to explore the subtly different connotations of these terms, I will say something about their application in the cases I discuss. I will often use "resurrect" as a placeholder that is agnostic as between these concepts.

[2]   Kenneth Asch, "Rebuilding Dresden," *History Today* (1999): p. 4.

the 1930s and those who engineered its reconstruction called it "a masterpiece."[3] Contemporary architects concur, mentioning its design and dome in the same breath as the design and dome of Michelangelo's Saint Peter's Basilica in Rome and the Cathedral in Florence, Santa Maria del Fiore. Put simply, the Frauenkirche was a grand and important piece of cultural heritage—not just for Germany but for Europe and indeed the world. Thus J. Paul, hitting some of these same notes, said of the Frauenkiche: "a special instance ... a curiosity with which there is really nothing to compare, and in the individual uniqueness of its architectonic solution ... the most original Protestant church building."[4] To borrow a term from Christianity: *amen*. The loss of the Frauenkirche was grievous and real. Dresden suffered mightily when the building toppled; so too did humankind.[5]

Plans for the Frauenkirche's resurrection started taking shape on February 15, 1945. And in the minds of some these plans never died. Indeed, during the nearly half-century the church lay in ruins, locals jealously guarded the site, preventing it from being cleared and safeguarding the church's remains. The fall of Berlin's wall acted as the catalyst to revive plans which had long lain dormant. Of course, not surprisingly—Dresden is a many-layered place with a complex history—controversy swirled around the various proposals. But reconstruction had wide support—80 percent of Dresden's City Council approved the project in February 1992—and in the end the decision was made to re-implement Bähr's blueprint. Indeed, the original plan was updated to reflect findings in the various technical studies—careful architectural/engineering surveys and materials analyses—completed prior to and after the destruction. This "new" plan did not modify Bähr's blueprint. Rather, and interestingly, it made more precise his implicit mathematical and basic structural intentions. (Bähr's design work toward the building as well as its remains were closely vetted and carefully analyzed by innumerable engineers and materials scientists; several authors note that the Frauenkirche is among the "most studied" buildings in history.[6]) The actual construction of the church—again, overseen by Bähr, stick in hand, himself—had veered somewhat from the original plan. And so small but important infrastructural improvements—all of

---

[3]   Among many other examples of this claim, see F. Wenzel, "A Construction of Stone and Iron: Structural Concept for Reconstruction of the Dresden Frauenkirche," in W. Jäger and C. Brebbia (eds), *The Revival of Dresden* (Southampton: WIT Press, 2000): pp. 175–84; quote p. 179.

[4]   Paul is quoted (p. 2) in Ottfried Jordahn, "The Rebuilt Frauenkirche in Dresden," *Studia Liturgica* 36 (2006): pp. 1–16. Jordahn also cites others who note the significance of the building.

[5]   Britain, which led the bombing raid, recognized this. Coincidentally, the day before the attack the *Manchester Guardian* urged that Dresden's glory be spared: "Dresden, with the charm of its streets and the graciousness of its buildings, belongs to Europe ... We hope it is spared the worst." Quoted in Jonathan Steele, "The Night it Rained Fire," *The Guardian*, 2/9/1995, p. T2.

[6]   See, e.g., Wenzel, "A Construction of Stone and Iron," p. 179.

which were required by contemporary building codes, consistent with the original design, and completely invisible—effectively realized the structural possibilities and wonder inherent in Bähr's blueprint. In short, imperfections and oversights in the implementation of the original "intuitive" plan were addressed and the building's stability enhanced. The reconstructed building, a second coming of the Frauenkirche, *verified* the glory of Bähr's Baroque building techniques.[7] Following what was a de facto archaeological dig at the site—among other things, the church's 1738 altar, largely intact, was discovered under the rubble—work began. The "archaeological reconstruction," as the project was officially dubbed, employed the two standing sections of ruin as building blocks around which the church was reconstructed. Salvaged material—fragments were photographed, carefully evaluated, catalogued, and sorted—was used wherever possible and the remainder of the building was fashioned from the same type of sandstone. Indeed, with the aid of IBM's three-dimensional imaging program (CATIA), 8,425 original ashlar stones were carefully reintegrated in their original locations throughout the building.[8] Mark Jarzombek describes it well: "the decision was made to treat the stones as elements of a vast, DNA-like research puzzle. They were separated, measured, analyzed, and then placed into the fabric of the new walls of the church, hopefully at the very spot where they once belonged."[9] Thus the original material— darkened and battle-worn, and sewn like seeds of memory—was interlaced with new, lighter material. At present the two can be readily discriminated. But weathering will render them indistinguishable. Eventually the Frauenkirche's scars will fade.

A success? Did the project resurrect religious-cum-spiritual value? I believe the answer is "yes," but before going on to show how and why, I want to consider two arguments *against* this conclusion. The first urges that reconstruction failed because, strictly speaking, there was nothing to resurrect. The horror of February 15, 1945

---

[7]   See Wenzel, "A Construction of Stone and Iron." See also G. Zumpe, H. Rothert, and M. Lugenheim, "George Bähr's Constructional Concept and the Reconstruction of the Cupola of the Frauenkirche in Dresden," in Jäger and Brebbia, *The Revival of Dresden*, pp. 197–218. The authors distinguish between "Bähr's Construction" and "Bähr's Constructional Concept." They argue that reconstructing the latter brings to fruition Bähr's real intention and resuscitates a grand piece of architecture unique in the world. Finally, see W. Jäeger, H. Bergander and F. Pohle, "The Reconstruction of the Sandstone Cupola of the Frauenkirche in Dresden," in Jäger and Brebbia, *The Revival of Dresden*, pp. 219–36. This essay, too, discusses Bähr's plan. The authors urge that if not for the fire the building would have "lasted another 200 years." Again, structurally, it was magnificent.

[8]   Ironically, the program used to "virtually reconstruct" the church had been developed for military purposes. See http://www.frauenkirche-dresden.de/daten-fakten-aufbau+M5d637b1e38d.html for details regarding the original materials used in the reconstruction.

[9]   Mark Jarzombek, "Disguised Visibilities: Dresden/'Dresden'," in Elaine Bastéa (ed.), *Memory and Architecture* (Albuquerque: University of New Mexico Press, 2004): pp. 49–78.

notwithstanding, H.-J. Jaeger says "the church stayed alive in the imaginations and hearts of the people of Dresden."[10] What makes a church a church? The argument urges that the church *really is* its congregation. The being of the church is the being of, or association between, its members. And so the Frauenkirche, rightly understood, never went away. In short, the building is inessential. Its reconstruction was at best unnecessary and at worst quite unsuccessful. (The Frauenkirche's congregation had long since begun to attend other churches. And the resources devoted to the project might have been put to other worthy uses.)

This is suggestive. There is, indeed, something to the idea that "the church" continued to exist absent sandstone and mortar. The congregation is important and the building would not have been a church without them. Obviously I cannot fully explore the ontology of "churchness." But I believe the argument trades on a problematic and strained metaphysics, undervaluing the building per se. In one sense the building *gave rise* to the congregation, making a collective out of disparate individuals; its former material being, its very physicality, served to forge the congregation. Indeed, the building served as the locus of, even as a condition of, their association, giving the parishioners a collective, mutually shared identity. (Ideas—including the idea "we are a congregation"—supervene on material stuff; they do not come out of nowhere.) Thus there was a *synergy* between the congregation and its building—a synergy which explains people toiling to load rubble carts with the church's material remains, safeguarding these for future reconstruction, as well as old-timers claiming to "know where every stone belonged"—and the argument misses this. In my view the parishioners' tangible and deeply felt concern for and focus on the building—for its parts and its reconstruction—*shows* that the building was not inessential, but rather crucial, to the congregation and to "the church." Without a home the congregation would not have been "Frauenkirchers."[11] They could have gone on, for a time, without their building; but eventually the Frauenkirche's congregation would have dissipated, merging once-and-for-all into other congregations or falling away altogether, in the absence of their home. Working toward reconstruction forestalled this.

---

    [10]   Jaeger, H.-J., "The Citizens' Initiative to Promote the Rebuilding of the Frauenkirche in Dresden," in W. Jäger and C. Brebbia (eds), *The Revival of Dresden* (Southampton: WIT Press, 2000): p. 149.

    [11]   A reverie-filled comment from Dr. Karl-Ludwig Hoch, a retired pastor who witnessed the bombing raid, is telling: "For two hundred years my family has been connected to the Frauenkirche. My grandmother was confirmed there, my great grandmother baptized, and I had books and pictures recalling those years. I know the church from top to bottom. I was only a boy at the time but I loved the Frauenkirche. The day after the raid we stood on our balcony where we lived. The smoke lifted for a moment ... it was unbelievable. I shouted to my mother, 'look, the Frauenkirche is still standing.' We stared again, and saw through reddened eyes it was true. Yet shortly afterwards it collapsed, as if not wanting to survive the insanity. The disaster was now complete." Hoch's comment, cited in Asch, "Rebuilding Dresden," manifests the import of the building.

A second argument urges that reconstruction was misguided—and the project a failure—inasmuch as the *ruins themselves* were a religious icon. In 1966, the church's remains were officially declared an anti-war memorial. Thus one might argue that the site symbolized moral evil and barbarism; the Frauenkirche's remains were a powerful reminder of the dangers and horrors of warfare, and served as a counterbalancing religious ideal. Indeed, one could urge (some did) that the ruins had come to symbolize peace—and should have been preserved in the interest of promoting the same. Put simply, this argument has it that the project destroyed something with religious meaning and value. And thus reconstruction was misbegotten.[12]

I disagree. To be sure, the site had become a symbol. But of what, exactly? While in one sense the ruins had come to stand for peace, they could also be read— this did not take much imagination—as a monument to ideological conflict and destruction. Indeed, exploiting this, the GDR used the site, politically, against the West and its ideology. Thus it is essential to try to disentangle the site's religious and political symbolism. But leave that (challenging) task aside. Assume that the rubble stood, most fundamentally, as a reminder of the evil of war and of its antidote, peace. Well, surely the reconstructed Frauenkirche, too, stands for peace. Just as surely it *also* stands for hope, renewal, and rebirth—each of them powerful religious ideals—and the ruins did not do this. Moreover, its slowly vanishing scars powerfully symbolize the religious values of forgiveness and reconciliation. Thus if religious value was destroyed when the rubble was cleared, so too was religious value recreated when, and as, the building went back up. (Photographs of the reconstruction-in-progress and its final result—when juxtaposed with photos of the ruin—*show* that the project was a success. Sometimes a picture is worth a thousand words.[13]) Indeed, reconstruction *fostered* religious value. And the resurrected Frauenkirche harbors more such value than did the ruins. In short, then, this argument, like the first, seriously understates the import of and value in *the building*.

These are quickly sketched observations. More might be said, for and against, each of the preceding arguments. But I want now to note—more specifically, and from less to more abstract—some ways in which reconstruction resurrected the religious-cum-spiritual values the building had harbored. First, the building is used as it was pre-1945.[14] There are religious services, musical performances,

---

[12]   There are a number of works on the various types of meaning and value in ruins. Among the most interesting and useful are Robert Ginsberg, *The Aesthetics of Ruins* (New York: Rodopi Press, 2004) and Paul Zucker, *Fascination of Decay: Ruins: Relic-Symbol-Ornament* (Ridgewood: The Greg Press, 1968).

[13]   Jaeger is one source of such photographs. See, e.g., "The Citizens' Initiative," p. 156 and pp. 158–9.

[14]   This is one criterion of authenticity—and hence of a successful restoration. For a searching discussion of the concept, see *Nara Conference on Authenticity in Relation to World Heritage*, ed. Knut Larsen (Paris: UNESCO World Heritage Center et al., 1995). This

fellowship gatherings, and the like; the Frauenkirche is once again an active center of religious practice. Second, its resurrection paralleled and symbolized the revival of religion from its public banishment, and moribund state, in the GDR. The reconstruction was a tangible marker of the reawakening of religious consciousness—officially *verboten* for decades—and it showed that religion was once again legitimate. Moreover, and most fundamentally, the church's reconstruction resurrected—and actively promoted—a paradigmatically religious thing or value: a sense of community or "togetherness." Indeed, the congregation—where this now includes not just parishioners but Dresdeners more generally—was *re-forged* in and through the reconstruction. Frauenkirche-goers, common citizens, tradesmen, and professionals alike worked, shoulder to shoulder, to resurrect the church. Put simply, the project *unified* the community. And surely this, following Tolstoy, is one of the central aims of and goods in religion. Even if the Frauenkirche's immediate congregation was in some sense held in abeyance—it was certainly discombobulated and disorganized—while its building was in ruin, it was plainly *re-fortified* and bolstered in the process of reconstruction.[15] In short, reconstruction solidified and extended the congregation—cementing it to a degree, and magnifying it in ways that would not have happened absent the project.[16]

And of course the reconstruction project went way beyond the Frauenkirche's congregation and way beyond Dresdeners to include the world. Indeed, I believe that the project—a huge and complex undertaking from start to finish, involving people from all walks of life, all religions, and all over the globe—actually *enhanced* the broadly religious meanings and values that obtained in the pre-damaged church.[17] In this respect, ironically and counter-intuitively, February 15, 1945 made possible new religious meanings and values, or perhaps new levels of the same meanings and values, where these are understood in a wider, more expansive, and more universal way. The *process* of reconstruction circumscribed whole new orbits of people—creating "Frauenkirchers" or "Dresdeners" *in spirit*—in a figurative but powerful sense. Thus the spiritual end of humans-in-community was amplified many times over by the reconstruction. The project resurrected a house of God and hence the parochial values of Christianity. But more than this it conduced to a deeply catholic or cosmopolitan end. From the aged

---

volume explores and "problematizes" the concept of authenticity, examining its history and portraying its different dimensions.

[15]   Did the *same* congregation and the *same* church arise? Were the congregation and the church restored or recreated or replaced? These are interesting metaphysical questions I cannot explore here.

[16]   Parallel discussions in this volume—see, e.g., Chapters 3, 4, and 8—point to the ways in which restoration projects in the natural world, restoration ecology, give rise to ritual and community. See also William Jordan's *The Sunflower Forest* (Berkeley: University of California Press, 2003).

[17]   Andrew Curry notes that private donors from 26 countries contributed to the project. See Curry, "Crowning Glory," *Smithsonian* 36 (2006): pp. 92–4.

parishioners who loaded the rubble carts to donors from diverse faiths worldwide to the engineers and tradesmen who labored to reconstruct the Frauenkirche stone by stone, the project fostered community exponentially. Indeed, the "Call from Dresden" was heard loud and clear—and answered with a resounding "count me in." Worldwide, multitudes of people joined hands, literally and figuratively, to resurrect the Frauenkirche. And this manifests the resurrection of religious-cum-spiritual value par excellence.[18]

Bamiyan's Buddhas: Their History, Significance, and Potential Resurrection

Bamiyan's Buddhas were the quiet and regal guardians of Afghanistan's Bamiyan Valley—a lush and peaceful rest stop on the Silk Road, the famously important trading route along which ideas and goods once traveled between China, India, Central Asia, and the Mediterranean Sea. The sculptures, perched in their niches since the sixth century, modeled the eclectic mix of cultures that had migrated through the region. They were a complex amalgam of many diverse styles and influences—Persian, Indian, Hellenistic, and Roman, among others—and telling material representations of the history and art history of Gandhara Buddhism. Indeed, Bamiyan's sculptures—the largest likenesses of the Buddha in the world and, arguably, the centerpiece of Afghanistan's material culture—had borne witness to roughly 1,500 years of a rich and cosmopolitan history. Previous military campaigns—Alexander the Great, Genghis Khan, and, more recently, the Soviet Union marched through Bamiyan's Valley—had exacted a toll. Weather

---

[18]   I now want to register a worry. On the one hand, the scars' eventual fading—within 50 years the original sandstone will be visually indistinguishable from the new sandstone—powerfully symbolizes forgiveness and reconciliation. In this respect the project plainly conduced to a deeply religious or spiritual end. On the other hand, religion's success is history's failure. *The scars are important.* They represent history. And thus with historical value in mind, the scars might well have remained *forever.* This "tension" shows that and how restoration projects serve different masters—multiple and often conflicting values—and they frequently, maybe invariably, leave us with remainders. Thus I don't mean to suggest that the Dresden project was an unequivocal success. Indeed, as I noted earlier, the project has *other* dimensions as well. And some of these will cut against the successful resuscitation of religious-cum-spiritual value I have discussed. In fact there is a second worry, as it were, *within* religion. While I cannot take this up here, it is important to note that the Friends of Dresden also issued a call for the resurrection of the city's synagogue, destroyed on Kristallnacht in 1938. While Dresden now has a new home for its Jewish citizens, the resurrection of the synagogue has been given short shrift as against the resurrection of the Frauenkirche. Jarzombek, "Distinguished Visibilities," discusses this. So, too, does Susanne Vees-Gulani, "The Politics of New Beginnings: The Continued Exclusion of the Nazi Past in Dresden's Cityscape," in G. Rosenfeld and P. Jaskot (eds), *Beyond Berlin: Twelve German Cities Confront the Nazi Past* (Ann Arbor: University of Michigan Press, 2008): pp. 25–47.

and time had done real damage, too. But the Buddhas were for all that proud and dignified survivors—until recently.[19]

On February 26, 2001, citing its particular interpretation of Islam, the Taliban announced its intention to bring down Bamiyan's icons. Over the ensuing week, while the world scrambled to prevent this, Afghanistan's de facto government positioned ammunition and weaponry. And on March 3, much to the world's collective horror, it commenced with the demolition work. But the Buddhas, like the Frauenkirche, were hardy. In fact it took a sustained effort, with both artillery and explosive charges, for the sculptures to finally succumb. By March 12, however, the Taliban's barrage was complete and the Buddhas lay in ruins. Shell-shocked Bamiyaners immediately began sifting through rubble. They were drawn to salvage and safeguard the remains of their sculptures, and they were intent on resurrecting the same. Indeed, the local citizenry and Afghanis more generally seemed to follow a kind of unconscious but deeply felt imperative: Bamiyan's Buddhas would go back up.

The "will to resurrect" points to one similarity between Dresden and Bamiyan. I will note some others—this is not an exhaustive list of reasons to think the cases analogous—without unpacking them. First, both pieces of material culture were locally iconic. Second, in each case the destruction was cruel and wanton.[20] Third, while both artifacts were created by flourishing and materially well-to-do cultures, the societies facing "the resurrection question" were materially poor. Fourth, precisely because of these hardscrabble conditions, both projects would require significant support, financial and otherwise, from other societies. Both projects would require the world to step up. Fifth, both artifacts—obviously—have a sacred element or dimension. Sixth, while the Frauenkirche and the Buddhas represent particular, parochial religious traditions, they also represent, more fundamentally, *world heritage*. Indeed, both were universally acknowledged cultural treasures. (In 2003, UNESCO added the Bamiyan Valley to its World Heritage List, simultaneously noting—even if, sadly, this was all-too-late for the sculptures—its endangered status. Dresden's Elbe Valley was added to the same list in 2004.) Seventh, a Frauenkirche-style project is *the type* of project Bamiyaners would undertake. Just as Bähr's original plans were excavated from the bowels of the church and employed in the reconstruction, careful analytical studies, materials

---

[19]   A useful source on the sculptures—their history and the history of Bamiyan more generally—is *Bamiyan: Challenge to World Heritage*, ed. K. Warikoo (New Delhi: Bhavana Books & Prints, 2002). This collection followed a September 2001 international conference organized to discuss the events of the prior March.

[20]   While there is some controversy about this, most historians agree that the Dresden bombing was an unjustified destruction of culture, serving no essential military purpose. (Of course the bombing also took thousands of lives. But here I'm focused on material culture. People cannot be restored.) And most view the Taliban's act as similarly gratuitous. While there are things to be said on the other side—the Taliban's motive is disputed—most find the desecration at Bamiyan utterly unconscionable.

analyses, and photographic documentation that had been completed on the Buddhas would be used in Bamiyan. In both cases we have the relevant blueprint; neither project would be "conjectural." Eighth, just as substantial original material was available for use in the Dresden case, so also at Bamiyan. In fact the recent rubble-clearing dig at the site indicates that "substantially all" the original material survived, much of it in useable form.[21] (Thus with useable original material—a central criterion of authenticity—in mind, the Bamiyan case perhaps holds out even more promise than did the Dresden case. The resurrection might be best understood as a restoration rather than a reconstruction or recreation.) Ninth, resurrected Buddhas, like the Frauenkirche, almost certainly would have a "motley" appearance. They would wear their history—the "good" history of Buddhism and the "bad" history of the Taliban's violent desecration—on their sleeves.[22]

Of course Dresdeners—understood here in a wide sense to include all those who contributed to the project—got their wish. Resolve became reality; effort paid off; the iconic church went back up. And I have argued that in terms of resurrecting religious-cum-spiritual values, the reconstruction of the Frauenkirche was a resounding success. Were the values in question *recreated* or *restored* or *replaced*? There are certainly important connotative differences between these terms. But in my view there is also a real sense—at least where the focus is on, as it has been here, religious value—in which it does not matter.[23] The church is once again standing. Did it expire, only to be raised from the dead? Did it take a long nap? Or was it reincarnated in a new form? To be sure, these questions point to and mark off interestingly different metaphysical possibilities. But however we are to understand the Frauenkirche's being, the meanings and values it harbors are *here*—they are back in the world, influencing us in good and important ways.[24]

---

[21]   See Carlotta Gall, "From Ruins of Afghan Buddhas, a History Grows," *New York Times*, 12/6/06, www.nytimes.com/2006/12/06/world/asia/06budd.html. Gall describes the current state of play at Bamiyan. The fragments are now under cover, awaiting a decision.

[22]   And unlike the Frauenkirche they could do this indefinitely, even forever. Thus history would be the winner, even if other value dimensions—like a purely aesthetic focus on appearance, for example—would not. The project need not be historically deceptive or misleading, as a pristine reconstruction might be—and as, once again, the Frauenkirche arguably will come to be. In short, the project could be carried out in a way such that history would not be erased.

[23]   As I have urged, the resurrection of religious value is manifest in community—and this can be resuscitated, even developed and enhanced, irrespective of the mode of resurrection. By contrast, the mode of resurrection *would* matter with regard to, say, historical value. Restoration would resuscitate the latter. Recreation would not.

[24]   And surely this would not be so, at least not to the same degree, if the site had remained a ruin. The religious value in the ruins—the sense of solidarity and togetherness to which they gave rise—was centered on loss. And preserving the ruins—as against reconstructing the church—would have permanently enshrined a lugubrious consciousness and mood. By contrast, again, the reconstructed church is a powerful symbol of hope, renewal, and reconciliation.

And now there is Bamiyan—a more recent and similarly appalling case of cultural barbarism. I have listed some striking parallels between these two cases. So the question arises: does Dresden's experience give us a good model for what ought to happen at Bamiyan? Should the Buddhas be resurrected? If so, how?

Any number of proposals—from leaving the niches empty to recreation to restoration to replacing the sculptures with laser mock-ups—have been floated.[25] In my view the answer, though certainly not non-controversial and not in any sense simple, is nonetheless plain: Bamiyan's Buddhas should be, in part because they can be, restored.[26] Indeed, Dresden's resurrection project, some six decades in the making, is an example of the sort of successful project that awaits Bamiyan. Resurrecting the Buddhas promises to resuscitate the various types of meaning and value that, lamentably, were rendered inaccessible in March 2001. Restoring Afghanistan's icons stands to rekindle aesthetic-cum-historical value; it also stands to contribute to each of economic and political value, making practical, real-world contributions to the lives of the locals who desperately want their sculptures resurrected. Moreover—to gesture now toward what has been the focus of my chapter—the *process* of restoring the Buddhas stands to give rise to the same sort of fundamentally religious-cum-spiritual value that arose in and through the reconstruction of the Frauenkirche. It stands, that is, to forge Bamiyaners, as well as both Afghanis more generally and the rest of the world, in the solidarity of a common project—and hence in community.[27] Indeed, as I have shown, the Dresden project extended the Frauenkirche's congregation to all corners of the globe. ("We are all Dresdeners.") And I think something similar is possible at Bamiyan. A restoration project could conduce to the same sort of spiritual end, linking Bamiyaners and Afghanis (a largely forgotten people, unless we're focusing on corruption, dissension, and violence) with the rest of the world. Ironically, then, the Taliban's destructive act—while plainly disastrous from the perspective of art, culture, and history—could still have this positive implication.

---

[25]   In a work-in-progress, "Should the Bamiyan Buddhas Rest in Peace?" I explore arguments for leaving the site "as is."

[26]   See my "Bringing Back Bamiyan's Buddhas," *Journal of Applied Philosophy* 28.1 (2011): pp. 44–64, for some of the complications involved. By "restoration" I intend the reassembling of original material. Experts have urged that anastolysis, the technical term for this sort of restoration, while obviously a Herculean task, is in principle possible. See, e.g., the Seventh UNESCO Expert Working Group on the Preservation of the Cultural Landscape and Archaeological Remains of the Bamiyan Valley (2008): http://whc.unesco. org/uploads/events/documents/event-563-1.pdf.

[27]   Just as the West helped bankroll Dresden's project, obviously this would need to happen at Bamiyan as well. Perhaps wealthy Muslim societies would pony up—as restitution for the Taliban's (or Al-Qaeda's?) idiosyncratic interpretation of Islam; as restitution toward Buddhist societies; as a sincere nod toward toleration and coexistence with other world religions; and as a signal to the world that Islam is not "crazy." This kind of contribution would conduce to the world's "togetherness quotient," just as it did in Dresden.

It could yet be the sort of unifying force, or spiritual conduit, that the Frauenkirche was in and for Dresden.[28]

Conclusion

Some final thoughts. One way of thinking about what should be conserved or preserved—and, where this is possible, as evidently it is in this case, *restored* and *resurrected*—is to ask the question: does the head recoil and the heart sink when we contemplate a world without the artifact in question? Dieter Hoffmann-Axthelm, discussing the conservation of buildings, says we should save only those structures "without which we would be poorer and the world would be cooler." He goes on to urge we should save only those buildings—his point applies, more generally, to pieces of world heritage like Bamiyan's Buddhas—"whose demise would break one's heart."[29] (His point also applies, quite obviously, to the ecological restoration projects discussed elsewhere in this volume. They have a similar motivation.)

Dresdeners (understood widely) thought and felt about the Frauenkirche. And now it is up to us to think and feel about the Buddhas. What do our heads and hearts say about Bamiyan? Should the Buddhas be saved? If we substitute "restore" for "conserve" or "preserve"—and, again, experts suggest that restoring the sculptures is feasible—I believe we see and feel immediately that the answer is "yes." In March 2001, Bamiyan, Afghanistan, and the world incurred a real and profound loss; the empty niches represent an open wound and a grievous harm.[30] But perhaps hope—again, a paradigmatically religious ideal—is not entirely

---

[28]   Here I want to briefly note and deflect a potential objection. One might worry that restoring the sculptures would perhaps resuscitate Buddhism at Bamiyan. While this is worth investigating, prima facie I think the objection is (way) off the mark. Why? Islam supplanted Buddhism at Bamiyan in the ninth century. And so the site had been devoid of "active" religious significance, for Buddhists, for well over a millennium. Indeed, in an important sense Buddhism "abandoned" Bamiyan long, long ago. And thus, as I have urged, restoring the sculptures stands to give rise to aesthetic-cum-historical value—as well as both economic and political value. Restoration would not resuscitate an active, practicing Buddhist site but rather a world heritage site. And it would conduce not to specifically Buddhist religious value, understood parochially, but rather to the more general and cosmopolitan and spiritual value I have discussed.

[29]   Hoffman-Axthelm is quoted in the context of a discussion about the best approach to Germany's cultural heritage. See Cornelius Holotorf, "What Does Not Move Any Hearts— Why Should It Be Saved? The *Denkmalpflegediskussion* in Germany," *International Journal of Cultural Property* 14 (2007): p. 37. If we apply this criterion to the Frauenkirche, I believe Dresdeners made the correct decision.

[30]   Some of the videos at the Bamiyan Community Development Portal site are suggestive here. See, e.g., www.bamiyan-development.org/art-and-media/al-jazeera-programme-on-youtube/. They powerfully portray the loss—and objectively articulate the dilemma Afghanistan faces.

lost. Discussing the decision to resurrect Dresden's church, Jarzombek says: "Overnight, the Frauenkirche became the symbol of the city's past, its survival, and its rebirth."[31] It seems to me a decision to restore the Buddhas might give us reason to hope for something similar for Bamiyan—and even for the troubled country of which it is a part. Broken hearts were mended in Dresden. And they might yet be mended in Bamiyan too.

Bibliography

Asch, Kenneth, "Rebuilding Dresden," *History Today* 49 (1999): pp. 3–4.

Curry, Andrew, "Crowning Glory," *Smithsonian* 36 (2006): pp. 92–4.

Gall, Carlotta, "From Ruins of Afghan Buddhas, a History Grows," *New York Times*, 12/6/06; www.nytimes.com/2006/12/06/world/asia/06budd.html.

Ginsberg, Robert, *The Aesthetics of Ruins* (New York: Rodopi Press, 2004).

Holotorf, Cornelius, "What Does Not Move Any Hearts—Why Should It Be Saved? The *Denkmalpflegediskussion* in Germany," *International Journal of Cultural Property* 14 (2007): pp. 33–55.

Jaeger, H.-J., "The Citizens' Initiative to Promote the Rebuilding of the Frauenkirche in Dresden," in W. Jäger and C. Brebbia (eds), *The Revival of Dresden* (Southampton: WIT Press, 2000): pp. 145–64.

Jäger, W., "A Short Summary of the History of the Frauenkirche in Dresden," *Construction and Building Materials* 17 (2003): pp. 641–9.

Jäger, W. and Brebbia, C., (eds), *The Revival of Dresden* (Southampton: WIT Press, 2000).

Jäger, W., Bergander, H., and Pohle, F., "The Reconstruction of the Sandstone Cupola of the Frauenkirche in Dresden," in W. Jäger and C. Brebbia (eds), *The Revival of Dresden* (Southampton: WIT Press, 2000): pp. 219–36.

Janowski, James, "Bringing Back Bamiyan's Buddhas," *Journal of Applied Philosophy* 28.1 (2011): pp. 44–64.

Jarzombek, Mark, "Disguised Visibilities: Dresden/'Dresden'," in Elaine Bastéa (ed.), *Memory and Architecture* (Albuquerque: University of New Mexico Press, 2004): pp. 49–78.

Jordahn, Ottfried, "The Rebuilt Frauenkirche in Dresden," *Studia Liturgica* 36 (2006): pp. 1–16.

Jordan, William, *The Sunflower Forest* (Berkeley: University of California Press, 2003).

Larsen, Knut, ed., *Nara Conference on Authenticity in Relation to World Heritage* (Paris: UNESCO World Heritage Center et al., 1995).

Seventh UNESCO Expert Working Group on the Preservation of the Cultural Landscape and Archaeological Remains of the Bamiyan Valley (2008). http://whc.unesco.org/uploads/events/documents/event-563-1.pdf.

---

[31]   Jarzombek, "Disguised Visibilities," p. 55.

Steele, Jonathan, "The Night it Rained Fire," *The Guardian*, 2/9/1995, p. T2.

Vees-Gulani, Susanne, "From Frankfurt's Goethehaus to Dresden's Frauenkirche: Architecture, German Identity, and Historical Memory after 1945," *The Germanic Review* 80 (2005): pp. 143–63.

Vees-Gulani, Susanne, "The Politics of New Beginnings: The Continued Exclusion of the Nazi Past in Dresden's Cityscape," in G. Rosenfeld and P. Jaskot (eds), *Beyond Berlin: Twelve German Cities Confront the Nazi Past* (Ann Arbor: University of Michigan Press, 2008): pp. 25–47.

Warikoo, K. (ed.), *Bamiyan: Challenge to World Heritage* (New Delhi: Bhavana Books & Prints, 2002).

Wenzel, F., "A Construction of Stone and Iron: Structural Concept for Reconstruction of the Dresden Frauenkirche," in W. Jäger and C. Brebbia (eds), *The Revival of Dresden* (Southampton: WIT Press, 2000): pp. 175–84.

Zucker, Paul, *Fascination of Decay: Ruins: Relic-Symbol-Ornament* (Ridgewood: The Greg Press, 1968).

Zumpe, G., Rothert, H. and Lugenheim, M., "George Bähr's Constructional Concept and the Reconstruction of the Cupola of the Frauenkirche in Dresden," in W. Jäger and C. Brebbia (eds), *The Revival of Dresden* (Southampton: WIT Press, 2000): pp. 197–218.

Chapter 8

Chanting the Birds Home:
Restoring the Spirit, Restoring the Land

A. James Wohlpart

I meet my students at the tennis courts on the campus of Florida Gulf Coast University at 7:32 a.m. on a Saturday morning. I am only two minutes late, yet I have already had calls on my cellphone from students wondering where I am. The sun has not yet risen, and the air is cool. The first front of the season finally pushed through southwest Florida, giving us clear skies and nighttime temperatures in the mid fifties.

I am with Sasha, my wife and an environmental educator at Florida Gulf Coast University; Dave Graff, an education specialist at Rookery Bay National Estuarine Research Preserve; and Chris Panko, a researcher also at Rookery Bay. We will meet Kim Mohlenhoff later; Kim is the exhibits coordinator at Gumbo Limbo Nature Center in Boca Raton. The five of us have become fast friends because of our mutual love of the natural world.

The students whom we are meeting are a mixture of first year students and alumni from the Learning Academy, a first year learning community at Florida Gulf Coast University. Students in the Learning Academy take integrated and interdisciplinary classes in cohorts of 20, increasing their engagement with each other, with faculty, and with learning. Retention and four year graduation rates of these students is higher than the general population, but because the program is costly, it has been cut as a result of a shortfall in state revenues. The housing boom slows, hurricanes miss Florida, the economy falters, and the effects ripple across the state, affecting old and young alike.

On this last trip of the Learning Academy to Riverwoods, the field laboratory for the restoration of the Kissimmee River, almost half of the 60 students originally signed up, but some of them fell away as the trip came closer—work, school, and trips to see family easily trumped this unknown adventure. We get oriented, signing waivers and reviewing maps, and we begin our caravan northward.

The drive of two hours takes us through some of the least developed land in south-central Florida. When we head north on Highway 29 out of LaBelle, we turn back in time. Along the way, we see a pair of crested caracara on the side of the road. I abruptly slow and pull over. The students are surprised at my quick stop, nearly causing a pile-up. Sasha, Dave, Chris, and I walk back along the road to get a closer look at the caracara; the students remain in their cars, heaters and music turned up.

A bit later, as we pass through the caladium fields of central Florida, we spot a lone pair of sandhill cranes. These majestic birds, which winter in Florida, mate for life and stay together year round, migrating south with their offspring in large flocks. They engage in an eerie ritual of unison calling, their heads thrown back, their beaks skyward. The female, who initiates the ceremony, utters two high-pitched calls; the male responds with a single call. With the longest fossil history of any extant bird, sandhills represent a steadfastness and commitment that is unusual in this short-term, short-sighted world. Their constant return to south Florida demonstrates a belief that the landscape that supports them will always be there to nurture them and their young. One year, when returning home from Riverwoods, we saw, in the newly plowed caladium fields, several hundred, perhaps even a few thousand cranes. For me they represent hope in a world gone awry.

One-hundred-year-old live oaks greet us when we arrive at the Riverwoods field laboratory, located on a 15-acre parcel of land owned by the South Florida Water Management District. The upper branches of these ancient giants appear to beseech the heavens in an enduring prayer; the lower branches bow down to Earth in quiet homage. Many are covered in Spanish moss that hangs from branches like John Muir's beard. Loisa Kerwin, the director, greets us. We gather for an orientation, where Loisa explains that the field laboratory is the home base for the restoration of the Kissimmee River, an ongoing project that began in the mid 1970s, and the site for environmental education about restoration.

In the 1960s, the 103 mile long winding Kissimmee River that ran from Lake Kissimmee to Lake Okeechobee was channelized, mostly a response to several years of serious flooding. The channelization, which created the 56 mile long C-38 canal, 300 feet wide and 30 feet deep, allowed for flood control and thus for cattle grazing, agriculture, and development.

Within a very short time of the completion of the project, the channelization proved to be trouble for the six million plus people who live in south Florida and the over 30 million tourists who visit every year. Without the floodplains where the water moved gently across the land, the aquifers were quickly depleted. Evidence accumulated as to the effects of the channelization on the health of Lake Okeechobee and thus on the Everglades. As a result of the runoff in areas with heavy cattle grazing, the phosphorous load tripled between 1970, when the channelization project was completed, and the middle of the 1980s. Phosphorous levels in Lake Okeechobee rose from 40 parts per billion to over 100. As goes the headwaters, so goes everything downstream.

The day ahead is an opportunity to explore the restoration project. Twenty-two miles of the C-38 are planned to be backfilled and two lock-and-dam structures will be removed, restoring 40 square miles of floodplain ecosystem, including 26,500 acres of wetlands and 43 miles of river. The cost for the original channelization of the river was 30 million dollars; the projected cost of the restoration is around 500 million dollars. Currently, nine miles of the canal have been backfilled, resulting in the water returning to its native flow—19 miles of meandering river.

The students will get to experience all three aspects of the Kissimmee River—a remnant run cut off from its traditional flow, the channelized canal, and the restored river.

We leave the field laboratory at 11:00 a.m. on the Kissimmee Explorer, a 30-foot pontoon boat. Captain Ken gives everyone safety instructions. We are on Riverwoods Run, a stagnant remnant of the original river. As we head downstream, cattle graze on either side. While the South Florida Water Management District has purchased all of the land in this area for phase two of the restoration, they have leased the land back to its original owners for cattle grazing until the planned restoration begins.

The students have all been given bird and plant guides, as well as a chart to tally the abundance and diversity of bird species. Loisa points out an osprey and a belted kingfisher. The students see an alligator floating in the water and then another, a juvenile with stripes still on its tail, lying on the bank. Laura, one of the students, confesses her strong fear of alligators, despite the fact that she grew up in Florida. When other students tell her that we have alligators on the FGCU campus, she refuses to believe them.

The historic floodplain, which extended up to three miles wide in some places, is bound by oak trees, which Loisa points to in the distance. This upland habitat, which may be only several inches higher than the floodplain, would have created the margins for the water that covered the area 70 percent of the year.

Loisa points to spadderdock, or cow lily, and pickerel weed, native plants, with a common gallinule hiding beneath. A tricolored heron lifts off the cow pasture, then floats next to the Kissimmee Explorer.

Cypress trees, natives to the broadleaf marsh, stand in the cow pasture like sentinels to a lost time, their needles gray, their knees no longer necessary because of the lack of flooding. Spanish moss hangs from their limbs. Sabal palms, our state tree, dot the landscape.

About halfway down the run, Loisa points to a limpkin on the bank and explains that this unique bird generally demonstrates the health of an ecosystem. Even though Riverwoods Run has not yet been restored, it may be reaping the effects of the restored sections upriver.

Matt, one of the first year students, points out a great blue heron, wading in the vegetation on the edge of the river. As we motor by, the heron lifts off, a pterodactyl from another era.

Students in the front of the boat point to two birds ahead, hovering over the water, chattering at each other. Belted kingfishers.

Two anhinga land in the top of a decaying cypress tree. Loisa explains that anhinga have an alternating flight pattern—flap, flap, flap, glide, flap, flap, flap, glide—while cormorants, which look very similar, have a constant flapping of wings.

Another great blue heron. Six American crows. The students tally the birds on their sheets, and one student keeps a master list.

Riverwoods Run will be part of the second phase of the restoration project, which will begin in 2012. The run is nearly stagnant, with a great deal of detritus settled on the bottom. The sand banks that accumulate in the bends of a flowing river are gone, which reduces the fish population that needs the sand banks for spawning.

The bahia grass that has been planted for the cows in the pasture will die off quickly when the water returns and overflows the banks. The native seeds, hibernating in the soil, will sprout and the cow pasture will become a broad leaf marsh once again.

Loisa points to rushes on the side of the run and asks students to see if they can spot any pinkish-whitish eggs. Ashlee—with a double "e"—sees them, pointing them out to the group. Loisa explains that they are apple snail eggs. By some miracle of evolution, apple snails lay their eggs at a different height on the rushes every year depending on how high the water will rise.

Crystal spots an alligator, and then a second on the bank, sunning.

Loisa points to a little blue heron and a tricolored heron, asking the students to see if they can tell any difference in their coloration. Christie notices that the little blue is fully blue, while the tricolored has a white breast.

A great blue lifts off and flies over the cow pasture. Matt and Laura call out, naming it.

Two white ibis on the banks are chased off by cows running from the sound of the boat's engines. We pass three cows in the river.

Tree swallows flit by overhead.

Another osprey, that students first identify as an eagle.

Ashley—with a "y"—points to a white bird, which the students misidentify as an ibis. A great egret, a stunning and majestic white against the green land. The grand bird lifts off lazily into the blue sky, the only white on a cloudless day.

A little blue, a great blue, another alligator hiding in the spatterdock with its heart-shaped leaf, and the cattails and water lettuce, an invasive exotic, and pennywort.

Another osprey, which the students name. Loisa explains that the osprey can be found on every continent except Antarctica.

One last belted kingfisher and we exit Riverwoods Run and enter the C-38.

The channelization of the Kissimmee River was an engineering masterpiece. In the 1960s, the Army Corps of Engineers created the C-38 canal, draining 34,600 acres of floodplain habitat and cutting off 68 miles of the once meandering river from historic water flow. The 56 mile long canal, with its lock-and-dam structures, protected residents from the regular flooding that was a part of the ecosystem, allowing them to use the land for grazing cattle and growing crops. With the removal of the water, species not native to a wetland habitat moved in and took over. Grass was planted for cattle. Fences went up. What was originally wild became managed, tamed. Humans moved in.

By the mid 1970s, locals recognized that with the changes in the river the way of life that had been handed down to them through several generations was gone.

No longer could they fish in the river, for the fish could not spawn in the canal. The birds that had graced their lives during the winter months no longer returned to visit. The historic river supported over 35 species of fish, 16 species of waterfowl, and 16 species of wading birds, along with river otters, alligators, and many other inhabitants of a wetland ecosystem, including an abundance of invertebrates, the basis of the food chain. With the invertebrates and the fish gone, the birds that once migrated through the region stopped coming.

Scientific studies confirmed what the locals knew from their own experiences. Migratory waterfowl had declined 92 percent. The bald eagle population had dropped 74 percent.

These studies eventually resulted in the 1976 Kissimmee River Restoration Act. Through this legislation, state and federal agencies began to work toward restoring the integrity of the river.

Then, in 1992, Congress authorized the Water Resources Development Act to begin the Kissimmee River restoration project, the largest ever of its kind. Scientists and planners considered a variety of possible restoration methods and developed models to ascertain what would happen as a result of the restoration. Finally, in 1999, the first phase of the restoration began with the backfilling of the canal and the removal of the first of two locks—the S-65B. The second phase of restoration, originally scheduled to begin five years later, has been postponed until 2012 because of an extended drought.

The students quickly note the difference between the remnant run and the canal. The native plants are nearly gone, the banks filled with Brazilian pepper, an exotic brought to south Florida for its seasonal, festive red berries.

With the plants and fish gone, the birds are also gone. The predominant bird seen during our trip on the canal is the cattle egret, which expanded into this area with the growth of the cattle industry. Originally from Africa and Asia, the cattle egret arrived in South America at the end of the nineteenth century and then migrated into the United States in the middle of the twentieth century. They are opportunistic feeders, following large machines or large animals like cows that tear up the earth and stir up insects, so they have thrived here.

The spoil mounds from the dredging of the canal, some 75–100 feet high, act like barriers on the landscape. Scientists have demonstrated that these mounds contain the seed banks of a bygone era. In the restored sections of the river, once the spoil mounds were pushed back into the canal, native plants quickly reasserted themselves.

Students spot a turkey vulture in the field. A little blue and a great blue fly over head. An anhinga pops his head above the water. Another sits in a Brazilian pepper, drying its wings. Crystal points to the oak trees that we had seen from a distance.

We enter the S-65C lock in order to lock up the river. Many of the students have never been through a lock. They are amazed at the large doors that close us in and the rush of water from the upper doors as the lock is filled with water. We start at 10 feet and rise to 18 feet, an eight foot lock up. Loisa explains that during the second phase of restoration this lock-and-dam structure will be removed.

As we leave the lock, we spy six cormorants resting on the buoys above the dam. A great blue heron hides in the bushes.

After a short run up the canal, Captain Ken cuts the engines. Loisa points up the channel where we see, 100 yards away, the end of the canal and the beginning of the restoration project. The former canal, now refilled with the spoils, has sprouted native plants in abundance. While not yet as rich or advanced in growth as the surrounding areas, the former canal is coming back to life. We turn into McArthur Run, the southernmost section of the first phase of the restoration project.

The abundance of plant life strikes the students immediately. Duck potato, with their white, spiked flowers, on the edges of the river. Wax myrtle, an upland native that moved into the area after the channelization, lines the river, dying now that its roots are submerged year round.

Red maple, in its glorious autumn colors. The students see a small heron and at first call it a little blue, but then notice the white breast and remember it is a tricolored.

Great blue. Turtles sunning on logs. Cypress trees line the river, vibrant. Unlike their graying counterparts in the cow pastures, these cypress trees have begun to turn orange, reconnected to their original habitat, to the land and water of the floodplain ecosystem, and to the turning of the seasons. Loisa explains that within a few weeks they will lose their needles and will remain barren until February, when they will green again.

Great blue heron, the students exclaim. White ibis. Loisa points to the pond apple growing on the edge of the river. Christie asks if they are edible. Dave suggests that they probably are. We all wonder what they taste like, though none of us dares to experiment.

A common moorhen emerges from the Salix willow. Old world climbing fern, an exotic, works its way up the willow, strangling it. Water lettuce and water hyacinth, exotics, are still in place. Occasionally they are sprayed with an herbicide to keep them in check or they will out-compete the natives.

A little blue heron lands on top of a cypress tree. Belted kingfisher, the students exclaim. Two great blues. Kim spots a snowy egret, the first we've seen, rising out of the water where it has been feeding. The students notice that it differs from the cattle egret that were so common on the canal, but now seem to have disappeared. The snowy egret, with its yellow feet, charms the students, a native returned home. Kim spots a lesser yellow legs. And then a marsh hawk with its white stripe across the tail.

As we wind our way through the restored river, the students notice fish jumping. Sand banks crowd the inside of every turn, a sign that the river's flow has returned to more natural levels. Between the naming of birds and especially the excitement of new species, a hush fills the boat. An energy seeps out of the land. There is a special feeling here, a feeling that originates from the restoration, the healing—an energy different from the feeling that comes from pristine wilderness.

This landscape reminds us, as humans, of our rightful place in the grand order for it signifies our humility, our recognition that we don't know everything, that we can't—and shouldn't—control every last acre.

Snowy egret. Lesser yellow legs. Cattails on point bars. Alligators, turtles.

Captain Ken cuts the engine. We float in the stillness for a moment, the blue sky above, the green landscape around. Loisa points to the north and to the south and asks the students how these areas look different. They seem flatter. The trees are small. There are no tall cypress trees.

"What was here, do you think?" she asks. Silence. Matthew tentatively offers, "The canal?" "Yes!" Loisa responds.

The spoil mounds that had been pushed into the canal, with their seed banks hanging on for over 30 years, have come alive with all manner of native plants. In this place especially, the healing that is central to this project is palpable. A hum emanates from the landscape.

We continue upriver, entering Micco Bluff Run. The students see turtles sunning themselves. An eight foot alligator splashes into the water. Laura points to it with delight. A black-crowned night heron flies nonchalantly past the boat. A red-shouldered hawk perches in a tree. Students call out. Belted kingfisher. Turkey vulture. Common moorhen. Snowy egret.

Anhinga. Cormorant.

Greater yellow legs.

White ibis.

Marsh hawk. Crested caracara.

The live oaks that moved toward the river's edge are losing their leaves and dying, falling into the river and creating snags.

Red-bellied woodpecker. Kestrel.

Dave points to a dozen or more golden orb weavers that have spun their webs across the tops of the trees. We slow so that he can take a picture.

Common moorhen. Snowy egret. Black vulture. White ibis.

We chant the names of the birds, we chant with the turning of Earth, with the healing of the land. Between the chanting there is silence, a quiet homage to the healing. It is as if we have entered an ancient cathedral. The live oaks that rim the floodplain honor the abundance and diversity of life in a broadleaf marsh ecosystem. They have been a constant reminder of the ancient wisdom of the land.

The stillness of the land enters us. We become settled. Our lives slow. The frantic busyness that characterizes our daily comings and goings is gone. We become present.

Great blue heron. Little blue. Tricolored.

Great egret. Snowy egret.

White ibis.

Anhinga. Cormorant.

Crested caracara. Osprey.

We chant the names of native birds returning home. We sense an ecology of the physical, an interdependent web of life that links the animate and the inanimate, which in turn creates an ecology of the spirit, a sense of the flowing energy that circulates through the universe, uniting us all. We remember that we are but one small part of a much grander system of air and water, plant and animal, rock and mineral, earth and sky and sun. We become like the oak trees, guardians, remembering our place.

We pass by a dead oak on the edge of the river. Dozens of golden orb weavers have made a home in the barren branches that reach to the sky. Kim explains that the larger female will eat the smaller male if she loses her web in order to have enough energy to build a new web. Life begets death begets life in a continuous cycle of renewal.

Nine miles of canal have been backfilled, creating 19 miles of wild, meandering river. While a true wilderness may never return to this place, a spirit of wildness nevertheless emanates from the healed land.

Wildness reconnects us to the wider relationships that weave across space like an intricate and fragile and sometimes invisible web. In our daily lives, we have become disassociated from these relationships and thus have forgotten our place in the wider order of the universe. We lose our humility. We act through arrogance and ignorance.

But occasionally, when we enter a sacred place like Riverwoods, we feel those connections, we unexpectedly run into the web that covers the path before us. We feel our entanglement and sense that our way of being can just as easily destroy the web as it can honor it.

In the restored river, we count 23 different species of birds, compared to the nine different species in the channelized river and 15 in the remnant run. The diversity and abundance of species offer a new lens through which we see the world. We have been living our lives in black and white, and all of a sudden it has exploded into a full spectrum of colors.

After eating lunch, we return down the restored river and down the channel. Our bird count done, the students sleep or sit quietly. Exhausted from more sunshine and fresh air than normally experienced, they rest. Those who sleep dream of crested caracaras and marsh hawks, birds that have now entered their imaginations and become a part of their very being.

When we enter Riverwoods Run, the students sit up, alone, quiet, connected. As we near the field laboratory, Crystal spots a limpkin on the bank with an apple snail shell in its beak. The limpkin gazes serenely at us, the aura of our encounter with the restored river emanating from our stillness. As we pass by, the limpkin nods as if in thanks.

The following morning, I wake early and run the short loop around the Riverwoods grounds. The sky is still dark, and I am alone, the students all sleeping comfortably inside. I begin my run, slowly, reveling in the experience of yesterday. As I run one loop, then another, and another, I connect with something ancient and primal. I fall into a meditative state, my breathing deep, a rhythm in tune with something bigger, the circling of days, the passing of seasons, the orbit of Earth. A fog rolls in.

The sun slowly rises, casting a glow on the Spanish moss hanging from immense live oaks. The sun shines off the dew heavy in the air. I circle the grounds, weaving a prayer of hope for this place of restoration and healing, losing myself in the wider energy of the universe.

Midway through my run, I hear sandhill cranes calling off in the distance, their marble throated call distinctive in the fog. First, the female calls in her high-pitched voice, two chants, then the male responds, a single chant. Back and forth they call until their chanting becomes indistinguishable, one. As I finish my run, the cranes glide through the fog and land in the center of the circle, before walking off into the fog.

I step inside and pour a cup of hot tea and then return outside to sit on the dock by the river. It is quiet again until the cranes call out somewhere in the fog. I hear them lift into the sky, and they circle overhead once, a wide circle, and then a second, much closer, a blessing bestowed upon us, a thanks for our prayers. They swoop low over the remnant run and disappear down the river behind a bend.

The second phase of the restoration has already begun with the mutual chanting of humans and birds. The sun rises higher in the sky, sending a million sparks dancing off the dew clinging to the air. I feel Earth turn under me, spinning through space.

PART 2
Recreating Place, Reconnecting with Others

Forrest Clingerman

For philosophy, ethics and theology, nature is a hermeneutical dilemma. What meaning and value can be gleaned from built and natural environments? Certainly such value cannot simply be utilitarian; human interactions with nature cannot be explained in light of what Heidegger called "standing reserve."[1] One significant way for theology and philosophy to approach a place and its inhabitants is as a text; places convey certain meanings that can be interpreted or read. The present section delves into some of the important conceptual issues that are implicitly raised when we begin to inquire into the meaning of environments in this way. Only a few of these chapters take an explicitly hermeneutical approach, but nonetheless what ties the chapters of this section together is what we might call an "environmental hermeneutical circle": landscapes and places are understood in the interrelationships between their inhabitants (rocks, trees, and woodchucks as much as humans). But those individual inhabitants are meaningful only in light of the whole place or landscape. Our interpretation of the environment is a local affair: we find nature when we approach individual places as texts.

Thus the next several chapters show some of the questions of interpretation that are at the heart of any philosophical or theological treatment of nature. As

---

[1]   In his seminal essay "The Question Concerning Technology," Heidegger explores the human relationship to nature as *Bestand*, which has been translated as "standing-reserve" in discussions of this essay. Julian Young, among others, suggests problems with this, preferring the more usual translation of the word as "stock" or "resource" (*Heidegger's Later Philosophy* (New York: Cambridge University Press, 2002), pp. 44–5). Through the term *Bestand*, Heidegger suggests that nature is not simply seen as raw material or stock in light of our technological world. Rather, it loses its place as an autonomous object entirely, and instead is subsumed in the aggressive framing of our world caused by modern technology. Nature is reduced to a part of the structure of mere usability, and thereby has lost any concealed, independent meaning. In sum, to say nature is "standing-reserve" is to suggest a more radical, instrumental sense than even a naïve utilitarian signification. It exists only insofar as it is ready to be used for our limited instrumental purposes. Cf. Martin Heidegger, *The Question Concerning Technology and Other Essays* (New York: Harper, 1977).

Martin Drenthen writes at the beginning of Chapter 9, "If we allow texts to open a world to us as a reader, then texts can change the way we understand ourselves and the world." But if interpretation is at the root of environmental thought, we cannot overlook the fact that hermeneutics is interested not only in *what* is being interpreted, but also *who* is interpreting and *how* one is affected by reading. Drenthen goes on to say, "If we want to take seriously the *meaning* of a place, we must also allow places to be of importance for the way we understand ourselves, as moral frames of reference, that can develop a much deeper relation with place, and a moral commitment with a place that goes far beyond merely recognizing a place's intrinsic value." Certainly, the complex ways we engage nature forces the authors of this section to consider more general concerns of how humans interpret—and continually reinterpret—the meaning of nature. Ecological restoration, as shown in the previous section, is an excellent example of this complexity. So how does our interpretation recreate places? What is our relationship with non-human beings, such that we find ourselves re-placed in meaningful environments? Where might we uncover the spiritual depth of place?

Since the present chapters are about how we encounter and understand the meaning of place and its inhabitants, the authors share a common desire to reflect on how our relationship with the others that share our environments "opens us to ways of being, to entire worlds, that we did not know existed" (in Anna Peterson's words). Clearly, what is being called for through the present volume is what David Utsler calls "environmental identity." And such a renewal includes a spiritual dimension, not merely an ethical one. Places become spiritual texts of creation, and traditional doctrines such as the Trinity are re-placed and envisioned anew.

Interpreting Places

The first three chapters of the present section respond to some of the ways that philosophy and theology can investigate not only our manipulation of place, but more importantly our relationship to ourselves and our environments in light of the places we inhabit. This section begins with "Reading Ourselves through the Land," by Martin Drenthen. Using an approach he explains as "environmental hermeneutics," Drenthen explores the idea of the "legible landscape" (taken from writer and landscape activist Willem van Toorn) as a hermeneutical tool for reading environments. In the Dutch setting, the concept of "the legible landscape" has been discussed by restorationists, and for Drenthen this term highlights how we gather together the meaning of places with the subject. He explains how the relationship of self and environment is defined as that between text and reader. We do not seek to simply gain information from reading landscapes as texts, and therefore a semiotic approach to such texts is insufficient. Instead, Drenthen argues the reading of environments opens us up to a critical and moral engagement with the world.

David Utsler's chapter, "Who am I, who are these People, and what is this Place? A Hermeneutic Account of the Self, Others, and Environments" complements Drenthen's use of environmental hermeneutics by showing how hermeneutics not only impacts our understanding of environments, but also offers new ways of understanding our "environmental identity." Drawing on the work of Paul Ricoeur on selfhood, Utsler suggests a narrative account of how selfhood is situated within environments. This environmental identity is found in the tension between distanciation and belonging. As with Drenthen's hermeneutical account, Utsler's investigation of self and place results in an ethical mandate. For both Drenthen and Utsler, the interpretation of landscapes and environments results in a narrative—and this narrative is a story about our sense of self and community. In a chapter entitled "Concern for Creation: A Religious Response to Ecological Conservation," David McDuffie suggests that "creation" is a point of mediation between ecology and Christian theology. It is not only a term for academic situations; he notes that it has liturgical and ethical resonance for the life of the religious community. Thus like Drenthen and Utsler, McDuffie argues that the task of interpreting the meaning of place—in this case, understanding non-human environments as divine creation—has a transformative effect on the self and the community. McDuffie might even be read as a theological response to Drenthen and Utsler: all three ask how to understand the deeper meaning of place and the transformative effect of environments on our sense of self and other.

Relating Self, Other, and the Divine

The next four chapters directly or indirectly approach place by reflecting on the interrelationships between inhabitants and environments. Together the authors are concerned with who or what dwells in place—and thus what is considered in environmental ethics and theology.

Anna Peterson, in "Recreate, Relate, Decenter: Environmental Ethics and Domestic Animals," focuses on a neglected topic: how our relationships with domestic animals can illuminate and broaden our understanding of the natural world. Domestic animals—particularly those animals that participate in our everyday lives and with whom we engage in play—resist the division between animal welfare and environmental ethics, Peterson argues. Play allows us to see the agency of non-humans and moves us beyond anthropocentrism. Yet this does not lead us to an undifferentiated affinity with non-humans: rather, it shows the need to integrate theory with concrete, lived practices. Partly in response to Peterson, Jonathan Parker's "Replacing Animal Rights and Liberation Theories" offers a different take on domestic and wild animals. While accepting that animal rights claims respond to the human relationship with domesticated animals, Parker argues that our relationship with wild animals requires a quite different ethic. In a world where "life feeds upon life," the case of predation is a clear example of the issues at stake in this ethic. In sum, Parker thinks we must partly replace animal rights with a great affirmation of wildness in our lives.

Peterson and Parker both investigate the relationship between human and non-human other. But can the divine also be found in place? Sarah Morice-Brubaker, in "Re-placing the Doctrine of the Trinity: Horizons, Violence, and Postmodern Christian Thought," serves as an example of how traditional theological doctrines are re-envisioned through discussions of place, and likewise how environments are re-interpreted through theology. Morice-Brubaker examines the theology of Jean-Luc Marion, in order to ask: where is God in this discussion of place? More specifically, Marion's Trinitarian philosophy seems to suggest a problematic relationship between God and spatial, embodied horizons. We are led to critique postmodern theology more generally, because it frequently situates the material as radically separate from an "unsituated" divine. There is a distance associated with the divine in postmetaphysical theology—a distance that we might argue is a radicalized parallel with the distance between human and non-human that both Peterson and Parker debate. Morice-Brubaker's critique, then, adds a further dimension to the discussion about the environmental hermeneutical circle of self and other, place and distance, human and non-human.

Finally, this volume ends with a "confession" that serves to move our discussion beyond life and spirit toward an even bigger whole. H. Peter Steeves' "In the Beginning and in the End" is a meditation on the relationships that exist between ethics, religion, and environments "from beginning to end." His approach is a fitting and poetic conclusion, and an excellent summary for this volume as a whole: rather than attempt to write a conceptual, dis-placed environmental ethic or ecological theology, Steeves ponders the intersections. He seeks an "ethic of all Creation," not just living nature. This is an ethic of place, meaning an ethic for rocks, trees, animals, and humans alike, for it asks: who are we? What is our place? And who is the other that shares this place with us?

Bibliography

Heidegger, Martin, *The Question Concerning Technology and Other Essays* (New York: Harper, 1977).

Young, Julian, *Heidegger's Later Philosophy* (New York: Cambridge University Press, 2002).

Chapter 9

Reading Ourselves through the Land: Landscape Hermeneutics and Ethics of Place

Martin Drenthen

Introduction

Philosophical hermeneutics is concerned with understanding the meaning and interpretation of texts. In each culture, certain texts are recognized as being particularly significant. Such texts present themselves as meaningful and help us understand what it means to be human. We feel there is something about these texts that has to be understood because we suspect they reveal something, e.g. about what it means to be mortal, or about the meaning of gratitude or compassion. In other words, such texts beckon to be interpreted, but in order to grasp the full meaning of such texts, we have to engage not just in reading but also in interpreting them. To truly understand the meaning of a text and what it has to say, one has to engage in the process of interpretation and let the words and letters of the texts bring forth a whole world. If we allow texts to open a world to us as a reader, then texts can change the way we understand ourselves and the world.

Environmental hermeneutics is concerned with the question of what hermeneutics can contribute to our understanding of environmental theory and practice. Environmental hermeneutics explore what it means to interpret environments, how environments can become meaningful to us, and how certain interpretations of the environment support certain self-interpretations. It is particularly interested in how specific places and landscapes present themselves to us as being significant and meaningful. We do not always already fully know *what* they have to say to us; but we feel their appeal on us: these places present themselves as significant and beckon to be understood and interpreted—"what is it about this place?"

In our daily lives, we usually do not find ourselves in mere abstract space, but rather in meaningful places. Our moral involvement with our surroundings is part of our being-in-the-world that roots in a conception of the world as an *ethos*, that is, as a morally structured, significant place for us as morally sensitive beings to live in. Therefore, according to Ingrid Stephanovic, "ethical discernment is less a matter of intellectual construction than it is one of attunement to a particular way of being-in-place. Rather than simply consisting of a project of internalizing an inventory of rules and principles, ethical awareness also unfolds prethematically and is informed by virtue of the ontological phenomenon of emplacement."[1] Such a view of ethics differs from more current forms of environmental ethics that tend to seek ethical guidelines for dealing with the environment in abstract notions such as "intrinsic value of nature" or "ecocentric egalitarianism." Such concepts are meant to help humans distance themselves as much as possible from their own anthropocentric partiality, "speciesist rationality" and "human chauvinism." From a hermeneutical perspective, such an approach on the human perspective is mistaken, because it presupposes a displaced and disembodied view of our being in the world, and—because interpretations about self and world are never innocent but always also determine how we lead our lives—runs the risk of actually transforming people into such abstract beings.

An environmental hermeneutical ethics does not start with a reflection and articulation of abstract values that people should adhere to. Rather, it starts out from the assumption that the world we live in has significance because it is always already infused with meanings. Moreover, hermeneuticists also stress that in order to grasp the full meaning of a particular place, one has to get involved in a process of interpretation. In that sense, landscapes can be compared with texts.

The idea that nature has to be interpreted in order to reveal a deeper meaning is not particularly original but has a long history. It has been first worked out in early Christian theology.[2] Augustine already said that God wrote two books: the Holy Bible and the Book of Nature. Reading the "Book of Nature" was not that different from reading the Bible: it enabled humans to feel part of God's creation and reflect on God's intentions with the world. The metaphor of the Book of Nature also played a role in eighteenth century German philosophy of nature, notably that of Goethe, who assumed that the workings of nature should be understood as revealing some deeper meaning of a World Spirit (*Weltgeist*).[3] These ideas do no longer seem to be of much importance in our present day, postmodern understanding of the landscape. Today, hardly anyone considers beauty in nature as a coded message from God, and romantic philosophy of nature has ceased to be of influence on natural sciences such as ecology and geology. Yet, in some contemporary ideas about landscapes, one can still sense some reminiscences of the old idea of nature as a text with a deeper meaning that must be interpreted.

One of these ideas is the "legible landscape," a concept that plays an important role in Dutch debates about landscape conservation. Typically, it is used to point out how old cultural landscapes—some more than others—contain signs that can be "read" as meaningful texts that tell a story about ourselves and our history,

---

[1]   Ingrid Stephanovic, *Safeguarding our Common Future: Rethinking Sustainable Development* (Albany: SUNY Press, 2000), p. 128.

[2]   Cf. Forrest Clingerman, "Reading the Book of Nature: A Hermeneutical Account of Nature for Philosophical Theology," *Worldviews: Global Religions, Culture, and Ecology* 13 (2009): pp. 72–91.

[3]   Cf. Henk Verhoog, "Lezen in het boek van de natuur," in Wouter Achterberg (ed.), *Natuur: uitbuiting of respect?* (Kampen: Kok Agora, 1989), pp. 141–59.

much in the same way as other texts from our cultural heritage do. By reading the landscape carefully, one can find a deeper meaning. The term was originally introduced by Willem van Toorn, for whom the ability to read a landscape is part of a much broader normative view on the moral relation between humans and landscapes. Yet, the term has been widely adopted by several conservation groups in the Netherlands, mostly because they believe that landscape legibility is the key to understanding—*and increasing*—people's attachment to particular places and landscapes. In 2004, the Dutch Association for Environmental Education (IVN) developed a method—*Project Legible Landscape*—to help local inhabitants increase their capacity to "read" the landscape and help them develop a deeper attachment with particular places.

In this chapter, I first briefly present Van Toorn's original concept. Next, I discuss the Legible Landscape Project and some of its underlying assumptions regarding the concept of landscape legibility. I show that the project is rather heavily influenced by a particular understanding of the relation between text and reader that focuses on revealing objective features in the landscape. In this perspective, reading a landscape boils down to getting to know the objective relationships between different recognizable features in a landscape. I argue that such a semiotic view of landscape reading is of limited value because it fails to recognize the moral implications of the concept of the legible landscape. Reading landscape texts does not merely revolve around information; landscape texts also present or "open up" a meaningful world to the reader and can thus play a crucial role in developing a view of self and world that can support an intimate and morally engaged relationship with these landscapes. I suggest that a hermeneutic account of the legible landscape is more suited for articulating such a normative vision of what it means to read a landscape.

Legible Landscapes as Food for the Spirit

The term "legible landscape" was coined by the Dutch poet, novelist and landscape activist Willem van Toorn. In his work, Van Toorn expresses his commitment and concern for the traditional Dutch river landscape, which plays an important part in both his novels[4] and essays.[5] In the late 1980s and early 1990s, Van Toorn joined a landscape protest movement that resisted plans of the Dutch Directorate-General for Public Works and Water Management to reinforce embankments along the major rivers (Rhine, Waal and Meuse). Due to both climate change and established water management policies, precipitation is expected to rise in the near future, leading the Dutch government to take precautions. For reasons of cost-effectiveness, it was decided not to strengthen the old dykes—often centuries-old,

---

[4]　Willem van Toorn, *Een leeg landschap* [*An Empty Landscape*] (Amsterdam: Querido, 1988); *De Rivier* [*The River*] (Amsterdam: Querido, 1999).

[5]　Willem van Toorn, *Leesbaar landschap* (Amsterdam: Querido, 1998).

small-scale, rather winding dykes—but replace them with higher, more robust and straighter ones.

The plan would be disastrous for the traditional typical Dutch river landscape, made famous by the works of seventeenth century landscape painters such as Jacob and Salomon Ruysdael, and would damage age-old remains of human habitation. The landscape activists warned that the implementation of the original plans would destroy many of these old "signs" and memory traces in the landscape (wooded banks, terraces, old small dykes and large artificial mounds), and would transform the "habitable, meaningful world" into a uniform, merely functional landscape dominated by straight lines; a "systematized" or even "medicalized" landscape (with artificial veins, bypasses and heart valves) devoid of meaning. Willem van Toorn was one of their spokespersons.

It was in this context that Van Toorn introduced the term "legible landscape": landscapes contain signs which enable people to "read" them as meaningful texts. Such landscapes "remind us along complicated and sometimes unconscious lines that there is a past, that people who lived in that past had to deal with the world just as we have to, that they had to protect themselves against nature and at the same time use its resources."[6] The reason we should value the legibility of the landscape has to do with our own sense of identity: "We have to stay in touch with this past— not because the past is better than the present, but simply because we owe our existence, our identity, our vision of the world to it, and because we can only think about the future by making use of our past experiences."[7] Thus, legible landscapes embody what O'Neill, Holland and Light call the "larger normative contexts" in which we can place our lives so as to have a sense of orientation; and that serve as frames of reference which tell us "what happens before us and what comes after."[8]

Eventually, and partly as a result of the landscape protests, the Dutch government decided that the dyke improvements should be integrated into the existing landscape with more care. At the same time, because in water management the dominant paradigm shifted from fighting against the water toward working with nature and giving room for the river, plans for dyke reinforcement also gave way to the ambition to restore the original river dynamics by "rewilding" large parts of the old agricultural land along the river. Yet, Van Toorn pointed out that these rewilding projects would have equally disastrous effects on the legibility of the land: again, legible signs that had been the result of a century-old history of human habitation would eventually disappear, this time wiped out by natural processes. Early protests warned against modernist civil engineering projects, now they opposed the ideas of green water managers, restoration biologists and urban wilderness lovers. In this new context, it became clear that the main worry about landscape legibility is anthropocentric, and concerns the meaning landscapes

---

[6]   *Ibid.*, p. 66.

[7]   *Ibid.*, p. 66.

[8]   John O'Neill, Alan Holland and Andrew Light, *Environmental Values* (New York: Routledge, 2008), p. 163.

have for human identity: "A landscape that does not contain enough 'signs', or where too many of these signs have disappeared, cannot tell us much."[9] From this perspective, landscape rewilding is just another modernist threat to landscape identity: "I consider it a dangerous development that, with nature construction, people aspire to create landscapes, in which humans are present only as tourists—and no longer as residents for whom the signs and narratives of the land are food for their spirit."[10] According to Van Toorn, rewilding will produce illegible landscapes, in which people will eventually lose their sense of identity and will become merely "visitors": tourists without a proper identity. What also becomes apparent is that for Van Toorn, the "legible landscape" is not merely a purely descriptive concept, but has a normative significance as well. In Van Toorn's view, being able to read a landscape is essential for having a meaningful and good life.

Van Toorn's concept of landscape legibility has received much public recognition. Apparently, it gave voice to a broadly recognized moral concern for the loss of significant feelings of place attachment. Moreover, it supported a vision of human life in which a meaningful existence should be embedded in a meaningful and non-trivial relationship with the landscape.

Many have attempted to acknowledge these kinds of concerns in environmental policy, by adopting the concept of landscape legibility in nature management and education. It is assumed that place attachment depends to a large extent on the accessibility of landscapes, both in the literal and epistemological sense. Landscape legibility is considered by many to be of key importance for increasing people's attachment to particular places and landscapes, which in turn is considered vital to assure future support for landscape conservation. Over the last few years, Dutch nature managers from the state forestry service and *Natuurmonumenten* (society for the preservation of nature monuments in the Netherlands) aim to increase the legibility of landscapes by protecting and highlighting particularly telling landscape elements. In addition, environmental education groups try to educate the public to recognize these legible features. Efforts have been made to translate Van Toorn's concept in such a way that they can be more easily applied in landscape governance policy. Yet as we will see below, in some of these translations, the normative aspect of Van Toorn's original idea tends to be overlooked.

Four Ways to Read the Landscape

In 2004, the Dutch Association for Environmental Education (IVN) adapted Van Toorn's idea and attempted to utilize the concept in a methodology for nature guided tours.[11] The basic idea is that nature guides organize short two-hour walks in

---

[9]   Van Toorn, *Leesbaar landschap*, p. 66.

[10]   *Ibid.*, p. 77.

[11]   Karina Hendriks and Henk Kloen, *IVN Handleiding leesbaar landschap* (Culemborg: CLM, 2007).

people's neighborhoods in which local residents and others are taught to "read the landscape." The purpose of these walks is to enable people to have a better/deeper *understanding* of a particular place, thus offering them a sense of orientation in space and time. By helping people discover the "stories that these landscapes tell," their relationship with these places is expected to deepen.

The IVN method distinguishes four different ways in which one can "read" a landscape. Together, they enable people to get a fairly complete understanding of the character, structure and meaning of a particular landscape.

The first perspective or manner of reading looks at the *vertical structure* of a landscape, and considers the relation between subsoil (soil composition, groundwater level, geomorphology, relief) and what grows and land use on the surface. Certain biotopes such as ditches, pools, hedges and wood banks can be used to determine soil composition (poor or rich in nutrients; backland or old river bank), which in turn can help us to understand which plant species should be at home here. Not all landscapes are equally legible in this vertical sense. Some places express the geomorphological structure clearly; here one can witness what happened in the (geological) past. Spatial patterns in these places tend to be rather complex, for instance because road patterns follow former river banks. In contrast, places which are structured merely on functional grounds are much less legible in this vertical sense. The square pattern of roads in the American Midwest, for instance, tend to conceal the structures underfoot and therefore lack verticality, in some sense at least. One could consider these landscapes are "equalized" and "flat." Contrary to what Van Toorn believes, ecological restoration *can* play a role in increasing the vertical legibility of the land, because some restoration projects actually highlight and uncover ancient, hidden texts such as old riverbeds and thus deepen the time horizon and add a longer sense of history in a certain place.[12]

The second perspective on the legible landscape focuses on the *horizontal structure*—those observable patterns that are visible on a topographical map. A horizontal reading gives us a clearer understanding of all the functional relationships in a landscape. It recognizes patterns of roads and waterways, zoning of housing areas, agricultural activity, ecological networks, but also whether an area is accessible for hiking and cycling; whether it plays a role in water storage or drainage, whether its spatial structure is open or closed, how different adjacent spaces are confined and separated from each other, etc. What do the spatial patterns tell us about functional, ecological and hydrological relations in this place? Is there a network of natural elements expressing the underlying ecological relationships? Landscapes with a high horizontal legibility tend to have complex,

---

[12]   An often-used method for rewilding flood plains in the Netherlands is the excavation of clay deposits, revealing the subterranean relief that preceded the cultivation of those flood plains by farmers. The Ark Nature Foundation even uses the economic yield of these clay excavations to pay for nature conservation activities, thus making large scale rewilding possible without much financial cost (cf. Martin Drenthen "Ecological Restoration and Place Attachment: Emplacing Nonplace?" *Environmental Values* 18 (2009): pp. 285–312).

multifaceted functional patterns and therefore tell an interesting story, whereas illegible landscapes tend to be monotonous and dull (e.g. the huge monocultures of the agro-industry).

The third reading of the landscape focuses on the *seasonal composition*: what do colors, shapes and structures in the landscape reveal about the particular time of year? In places with much variation throughout the year, one can use the colors and forms of plants, vegetation and crops to notice which season it is. In some places these changes are abrupt; in others they are more gradual. A poorly legible landscape looks the same all year round, whereas a highly legible place talks about the meaning of spring and fall, brings to mind meaning of the passing of time and conveys what impact the seasons can have on the land.

Finally, the fourth perspective looks at the *cultural history* of a landscape and how observable patterns and elements in the landscape reflect specific moments in history. Central to this perspective are anthropogenic landscape elements and patterns such as buildings and road patterns. Age, style, type of build and material (e.g. river clay) of buildings tell about the history of a place. A brick stone farm house on an artificial mound in a Dutch river area tells of a time when the river still moved freely and occasionally flooded the land, providing fertile soil and building material, and how humans adapted to that natural system. But the pattern of parcels of land can also reflect the organizational structure of society during times of land cultivation. Even vegetation can be an expression of specific times or phases in history: exotic tree species bring to mind the Romans who introduced them 2,000 years ago. Certain trees and bushes, which were planted by farmers to provide axe handles, still remain in the landscape long after the farm houses to which they belonged have disappeared. Finally, present land use (grass land, silt dump, clay excavation) tells us about the local historical phase that we are in, today. And all these signs can say something about the future prospects.

By teaching people to look at "their" landscapes through these perspectives, the Legible Landscape Project aims at making the land more intelligible by teaching people how to look. Together, these four ways of reading the landscape enable us to gain a complex understanding of a place and the story that it has to tell. Moreover, the project can also explain why some places appeal to us in particular: because they tell more interesting, more complex and more colorful stories than other places.

Like Van Toorn, although less clearly and more implicitly, the Legible Landscape Project also presupposes a particular normative ideal about how people and landscapes should relate. The project considers landscapes as texts worth reading, and tries to get people to pay closer attention to the landscape. Implicitly, it is assumed that a human life is more worthwhile in a landscape that is legible: the "spatial quality" or "landscape quality" is said to increase with increased legibility. But the project does not explicate *why* people should read the landscape more carefully, *why* it would make our lives worthwhile, and why it is better morally to read a landscape than—for instance—to watch a reality soap on television. To understand that, one needs a richer approach to landscape legibility.

The Semiotic Bias in Current Landscape Reading Practices

The approach of the IVN toward the legible landscape is heavily influenced by a particular conception of textuality—semiotics—that assumes that the meaning of a text derives from the connections between the constitutive parts within a text—the "signs." According to semiotics, understanding the meaning of a text, which is seen as a network of signs, primarily means that one can recognize these relations within this network and represent them in a network of symbols. For semiotics, anything can be read as a text, since all things in the world are part of a relational network with other things. With regard to the landscape as a text, the semiotic view focuses on the way in which individual elements in the landscape (a hedge, a farm, a ditch) refer to each other and form a network of interconnected signs.

It is this semiotic view of reading a text that is at the root of the four perspectives of the Legible Landscape Project, all of which refer to relationships between objective features that exist "out there." Reading the landscape in the semiotic sense merely focuses on recognizing all actually existing relationships of elements in the objective world. The focus is on knowing the relations between the legible elements: for example, recognizing how the species of trees is related to the soil composition, or how the road pattern reflects the way the river has shifted its course. Once I know the relationships between objective features of the world, I can represent these relations by telling the story of a place—a story that reflects the different complex relationships in the world in a symbolic order.

Because the semiotic approach *equates* reading a landscape with merely acquiring *information about the world* "out there," the semiotic idea of textuality is fundamentally unable to explain why people feel deeply connected and committed to landscapes. Its conceptualization of meaning starts with a clear disconnection between the objectivity of the text and an uninvolved reader, and thus presupposes (or rather creates) a gap between the understanding subject and the objective signs. This focus on knowledge as representation of a place ignores other ways of "knowing" a place, such as having an intimate acquaintance with a place. What semiotic landscape reading practices do not address is the relationship that *we* as readers have with this legible landscape. The semiotic reading shows why the story of certain particular landscapes may be considered more interesting than others, but it does not address why certain places appeal to us in a way that involves who we are. It does not tell us *why* people (should) care about the legibility of the land. Why they (should) want to get to know these places, why is there more "quality" in a legible landscape?

Thus, the semiotic view does not recognize the normative element that proved central to Van Toorn's concern: the "residents for whom the signs and narratives of the land are food for their spirit." The very connection between a legible landscape and a good life, which is at the heart of what Willem van Toorn originally tried to express with the concept, cannot be addressed through the type of landscape readings that the IVN project is promoting.

It should be acknowledged that the authors of the IVN method explicitly recognize that their four reading perspectives can only be a first step in the process of developing a deeper understanding of a place. Seeing and knowing the different aspects of the land can be the start of a conversation among those who made the guided tour to talk about what the landscape really means to them. But the IVN authors recognize that in our relation to place, other, more intimate meanings are in play. Landscapes also have subjective and personal significance for people; people develop a relationship with a landscape by connecting the objective story about the landscape with their personal experiences. The IVN project acknowledges that merely *knowing* a place is only part of the story. After the guided tour, the floor should be opened for a conversation about more personal experiences of the landscape. Yet because IVN's collective reading of the landscape starts from a semiotic conception of reading, the nature of these other meanings tends to be viewed as subjective and highly personal, and rather disconnected from the idea of nature as a text in need of interpretation. In the end, it is suggested, the landscape is different for everyone, since everyone has his own favorite spots, his own little stories to tell about the place. IVN acknowledges that the objectivist approach ignores the way that individuals experience and value particular landscapes. But to the extent that these more intimate meanings are conceived as being highly personal, it is difficult to see how they can play the normative role that they do in discussions about legibility of the landscape.

Like many who are unsatisfied with a merely objectivist account of the relation between humans and landscape, in order to address more personal meanings the IVN project reverts to environmental psychology as a subjective supplement. Environmental psychologists study how individuals experience and value certain landscapes and places; some even try to determine why people value particular places more than others. The problem with a psychological approach, however, is in the end that it too cannot explain the normative content of the reading of landscapes. For psychologists, experiencing meanings of a landscape can never be anything else but a mere subjective attribution to an otherwise value-neutral landscape; the landscape itself is merely the white screen on which individuals project their personal tastes and preferences. However, if all these other, more intimate and relational aspects of our understanding of place are merely subjective personal *attributes* of an otherwise "mute" objective landscape, then the experience of humans would not so much testify to something about the meaning of the landscape itself, but rather to the structure of their character and their personal history. Environmental psychology indeed complements the objectivist approach, but as a perspective on the legible landscape it is equally unsuited to understanding how certain places can present people with a moral frame of reference. Merely adding subjective experiences to objective legible features cannot explain how certain place meanings can support a morally deep sense of belonging in which places can be "food for the spirit." Both the semiotic focus on gathering knowledge about a landscape, and the psychological approach to nature experiences, remain deeply embedded in a notion of reading landscapes that is fundamentally unsuited to understanding how landscapes can serve a role as a normative frame of reference.

The semiotic perspective on the landscape presupposes a distanced perspective on place in which landscapes can at best be objects that tell interesting stories. We can choose to value these places, or even choose to recognize their intrinsic value, but a semiotic reading of the landscape would also be perfectly compatible with a detached, self-centered relationship with place. It is true that an average resident will know more about a place than an average incidental visitor will, but that does not mean that having more knowledge about a place will lead to an increased place attachment. Visitors can discover new information about a place, but they will at best become better informed visitors. A place can become more interesting to them, but that is not enough to support a true attachment to place and becoming a true resident. For those who are already attached to a place, the semiotic reading method can add depth and meaning to the landscape, but for those with a rather shallow relation to the landscape, visiting the place will at best be another nice thing to do on the weekend. And next weekend, they will visit another place that will be just as interesting. But was this precisely the type of tourist attitude that Willem van Toorn dreaded so much, when he complained about the prospect of becoming merely a "visitor to his own landscape"?

If we want to take seriously the *meaning* of a place, we must also allow places to be of importance for the way we understand ourselves, as moral frames of reference, that can develop a much deeper relation with place, and a moral commitment with a place that goes far beyond merely recognizing a place's intrinsic value.

Hermeneutics as an Alternative Perspective on Reading

Instead of starting with a semiotic conception of the legible landscape and supplementing that with purely subjective personal experiences, we should acknowledge from the start the intimate relationship that can exist between a text and a reader. In order to understand the actually existing moral ties between people and their landscapes, we should be open to a new ontology that focuses on the relation between "subject" and "object," people and their places. If we want to understand how the legibility of the land is connected with environmental identity and ethical commitment to place, we need a different perspective on legibility and textuality. Whereas semiotics tends to focus on texts as information carriers, hermeneutics tends to look at the way that interpretations play a part in our understanding of ourselves and the world.[13]

A hermeneutic approach toward the legible landscape can begin from the work of the French philosopher Paul Ricoeur (1913–2005) on texts and reading.

---

[13]   For this reason, hermeneuticists typically focus on literary rather than non-literary texts.

Ricoeur defines a text as "a discourse fixed by writing."[14] According to Ricoeur, there is an important difference between texts and speech. In speech, a speaker can accompany his signs and explain himself. Because in a text the author is absent, it is up to the reader to understand the meaning of the text through interpretation. When language is transformed into a *text*, Ricoeur argues, it assumes a life of its own, independent of that of its *author*; or phrased in a typically Ricoeurian manner: "the text has been emancipated from its author." If one wants to understand the meaning of a text, it can be inappropriate to ask the author what he intended to say in a particular text (even in the exceptional case that we *could* ask him), if only because some texts accommodate much richer readings than the author intended. In a way, the author is merely the first reader of the text, but has no privileged position to determine how the text should be read or what is the meaning of that text (this is especially true for literary texts). Without an external authoritative source to turn to regarding the meaning of a text, a reader can only revert to the act of reading the text itself.

Moreover, in speech a speaker can physically point to the things he is talking about; a speaker presents to an interlocutor a "real" world of which both speaker and interlocutor are a part. A text, in contrast, presents an imaginary world that has to be supplemented by the reader, if only because of gaps in the text's references, which ultimately must be filled by the imagination of the reader. "Texts speak of possible worlds and of possible ways of orienting oneself in these worlds."[15] But in order to understand the meaning of a text, we not only have to be open to the world as presented by the text, but we should also be willing to "place ourselves"—for the time being—in that world. To understand the meaning of a text means that we should not project our own beliefs and prejudices onto the text, but rather, that we "let the work and its world enlarge the horizon of the understanding which I have of my self."[16] Moreover, we never read a text in isolation—our understanding of the text presupposes the existence of preceding texts that have already helped to constitute both the reader and the world of the text as well.

Thus, text, world and reader are engaged in a dialectical relationship. According to Ricoeur, good reading requires willingness on the part of the reader to participate in the world that is opened up by the text and abstract from the context of one's particular life ("distantiation"). However, understanding a text also means to be involved, to be "present" in the act of reading. A reader has to bring to life the narrative of the text, bring to bear the meanings of words and concepts that play a role in his own life. Good reading therefore does not only require "distantiation,"

---

[14]   Paul Ricoeur, *Hermeneutics and the Human Sciences. Essays on Language, Action and Interpretation*, ed. John B. Thompson (Cambridge: Cambridge University Press, 1981), p. 146.

[15]   *Ibid.*, p. 177.

[16]   *Ibid.*, p. 178.

but also "appropriation": the reader must use the context of his own life to "bring to life" the world that is being brought forward by the text.[17]

Ultimately, text and reader are tied together in a narrative, in which the reader tries to explain the meaning of a text, but at the same time testifies to his own identity. It is through the texts that he reads and by imagining himself in the worlds that are being opened by these texts that the reader gets to know himself. Through the act of reading and interpreting texts, one gets to know "oneself as another."[18] It is this latter notion of identity that Ricoeur calls narrative identity.[19]

I believe that this Ricoeurian perspective can provide us with a new mode of understanding the legible landscape that enables us to understand how the legibility of a landscape can inform one's place-based identity and intensify one's relationship to such landscapes.

Landscape Hermeneutics and Ethics of Place

As we have seen, central to Ricoeur's hermeneutic conception of textuality is the idea that the meaning of a text depends on the act of reading, and that this reading act implies an active stance from the reader. If we want to connect this Ricoeurian perspective with environmental philosophy[20] and with the discourse on the legible landscape, we can start by noting that the legible landscape shares the features crucial for texts in a Ricoeurian sense: the legible landscape contains fixed signs that are in need of interpretation, while the author of this text is absent.

---

[17]   Elsewhere, I have shown that for us, today, there is also a significant limitation to this idea of appropriation, because in our postmodern age, we seem to have become too self-aware and too aware of the contingency of each particular appropriation of nature. Postmodern wilderness desire could be a symptom of this nihilistic self-awareness: we long for something that is not interpretation because we seem to lack a culture of nature—are not at ease in any cultivation of the world (cf. Martin Drenthen, "New Wilderness Landscapes as Moral Criticism: A Nietzschean Perspective on our Fascination with Wildness," *Ethical Perspectives: Journal of the European Ethics Network* 14 (2007): 371–403).

[18]   Paul Ricoeur, *Oneself as Another* (Chicago: University of Chicago Press, 1992).

[19]   David Utsler has repeatedly demonstrated that this Ricoeurian conception of narrative identity can be a model for understanding what Utsler calls "environmental identity" (David Utsler, "Paul Ricoeur's Hermeneutics as a Model for Environmental Philosophy," *Philosophy Today* 53 (2009): pp. 173–8).

[20]   In this section, I draw heavily on reflections on the relevance of Ricoeur for environmental philosophy by Forrest Clingerman on "emplacement" (Clingerman, "Beyond the Flowers and the Stones: 'Emplacement' and the Modeling of Nature," *Philosophy in the Contemporary World* 11 (2004): pp. 17–24), David Utsler on environmental identity (Utsler, "Paul Ricoeur's Hermeneutics") and Brian Treanor on "narrative environmental virtue ethics" (Brian Treanor, "Narrative Environmental Virtue Ethics: Phronesis without a Phronimos," *Environmental Ethics* 30 (2008): pp. 361–79).

According to hermeneutics, the act of reading a landscape presupposes that we are already "engaged" with a landscape that has already presented itself to us as meaningful and worth exploring. Landscapes are always already infused with meanings, embedded in a larger whole of meanings and interpretations that are already in play in how we see the world and ourselves. A hermeneutic perspective on landscape legibility thus starts with the connections and dialectics between text and reader that are always already at work. Instead of starting with a distinction between objective land signs and subjective experiences of meaning and value, and then having to face the question how to understand the connection between both separated entities, landscape hermeneuticists presuppose that there already exist several connections between landscapes that beckon to be interpreted, and readers who need to interpret their meaning.

Moreover, a landscape hermeneutic will not so much attempt to describe how certain groups and individuals happen to be interested in reading the land (as environmental sociologists and psychologists would), but rather seeks to show what it means that the landscape presents itself as a text worth reading.

According to Ricoeur, humans are truly narrative beings, who know themselves *through* the stories that are being told. Through the act of reading, a text can change both the reader's world and his identity. If the reader answers to the "invitation of the text," then the "refiguration of the world by the text" can bring about an active reorganization of the reader's being-in-the-world. Thus, according to Ricoeur, one's narrative identity is determined in part by the opening horizon of new worlds that are being disclosed by texts. Landscape texts and place-identities could be linked in a similar way as well. If landscapes and places can be read like texts, then the act of reading landscape texts could be formative for personal identity as well. Reading the land as text requires an active engagement with the meaning of a place that beckons to be articulated in our act of interpretation. We must therefore both actively appropriate the meaning by investing ourselves in the landscape, and at the same time let the text change our world. The stories that we tell about the meaning of a place and what it means to be in that place reflect and support our identity but can also transform it. Through the act of reading, the land can become intertwined with my own life story; it can tell me something about myself that I did not know before. Thus, it can become a true dwelling place—an ethos—that defines who I am and what my life is about.

Our interpretations of the world and ourselves have always already been partially constituted by earlier texts and interpretations in play. To take the analogy one step further, we can say that the same might be true with earlier landscape-texts: our current place-identity is partly determined by the way we have always already been emplaced. Our identity is already being formed by the place-narratives that surround us. We are always already "emplaced":[21] that is, we are being formed by the existing meanings and interpretations of the land.

---

[21] Cf. Clingerman, "Beyond the Flowers and the Stones" and "Reading the Book of Nature."

Insofar as this perspective makes sense, then understanding the legible landscape requires that we must also learn to understand how places have always already contributed to who we are. We should learn to understand ourselves *through the landscape that we find ourselves in*, and then move on to produce more adequate interpretations of the meaning of the land to enable more adequate practices.

The Role of Semiotics in Landscape Hermeneutics

It is at this point that the semiotic approach to landscapes might become especially relevant again. Hermeneutic and semiotic landscape reading could be complementary. An adequate place-based narrative ethics has to reflect and strengthen place-based identities, but also has to acknowledge that each place has a status, specific nature and history of its own. This "otherness" or "autonomy" of the world has to somehow be part of the place narrative as well, and semiotics could be a way to ensure that our stories actually reflect the nature of a place. Place narratives cannot be invented at will, but somehow have to be "grounded" in an understanding of the nature of this place: its history, its soil composition, the way it changes throughout the year, what species live there, what food you can grow there, etc. Our place narratives should consist both of stories about the nature of a place and about what it means to live here.

The semiotic approach brings into play the "objectivity" of a place by conveying certain place features with a specific "gravity." For Ricoeur, the world of the text is fundamentally different from the "real" world. A speaker can point to the things that he talks about—the world that both speaker and listener live in. A written text, in contrast, opens a world, but this world is incomplete, for it has many gaps which have to be supplemented by the imagination of the reader. Such a view could imply that the world that is being opened by the landscape text is in a sense "not real": it exists only insofar as it is interpreted. An old agricultural landscape can bring to mind long-gone worlds of traditional farming, where humans and land lived together in mutual dependence. Such a world only comes into existence by the active interpretative act from us—the readers and interpreters of this text— and yet these meanings are not freely invented, but result from a serious attempt to understand the meaning of the landscape. The narrative connects the land to a sense of our identity—the story of what it means to live on this land—but in addition the legible landscape also tells a story that is "real" in a much more literal sense. Forrest Clingerman has noted that, in this respect, the landscape is a very special kind of text, because the world brought forth by the landscape-text *is* the real world (in a very specific sense at least).[22] One could say that semiotics focuses

---

[22]   Cf. Clingerman, "Beyond the Flowers and the Stones," pp. 19ff., who also reflects on the differences and similarities between a Ricoeurian idea of emplotment and the materiality of a place-oriented notion of "emplacement."

the attention on these "real" features that the reader can "point out," and to the "reality" of the world of the landscape-text, in which the reader finds himself.

Thus, the IVN Legible Landscape Project can call our attention to how any particular place functions, how it has come to be, what intrusions it can and cannot take, etc. Yet, in the end, the objectivity of the semiotic approach has to be integrated with an overall hermeneutic of the landscape, in which all the objective features are put into context and get to *mean something*. It is this narrative context that connects our fate with the legible landscapes we live in. Only when the land is somehow already intertwined with our life story and narrative identity, can it provide an ethos—a normative context or frame of orientation with which we can orient ourselves and from which we know who we are and what being in our particular place is about. Yet objective landscape reading like that developed by the IVN provides a tool to become critical and reflective about the true meanings of places and landscapes and maybe adjust the image we have of ourselves.

Bibliography

Clingerman, Forrest, "Beyond the Flowers and the Stones: 'Emplacement' and the Modeling of Nature," *Philosophy in the Contemporary World* 11 (2004): pp. 17–24.

Clingerman, Forrest, "Reading the Book of Nature: A Hermeneutical Account of Nature for Philosophical Theology," *Worldviews: Global Religions, Culture, and Ecology* 13 (2009): pp. 72–91.

Drenthen, Martin, "New Wilderness Landscapes as Moral Criticism: A Nietzschean Perspective on our Fascination with Wildness," *Ethical Perspectives: Journal of the European Ethics Network* 14 (2007): pp. 371–403.

Drenthen, Martin, "Ecological Restoration and Place Attachment: Emplacing Nonplace?" *Environmental Values* 18 (2009): pp. 285–312.

Hendriks, Karina and Henk Kloen, *IVN Handleiding leesbaar landschap* (Culemborg: CLM, 2007).

O'Neill, John, Alan Holland and Andrew Light, *Environmental Values* (New York: Routledge, 2008).

Ricoeur, Paul, *Hermeneutics and the Human Sciences. Essays on Language, Action and Interpretation*, ed. John B. Thompson (Cambridge: Cambridge University Press, 1981).

Ricoeur, Paul, *Oneself as Another*, trans. Kathleen Blamey (Chicago: University of Chicago Press, 1992).

Stephanovic, Ingrid, *Safeguarding our Common Future. Rethinking Sustainable Development* (Albany: SUNY Press, 2000).

Treanor, Brian, "Narrative Environmental Virtue Ethics: Phronesis without a Phronimos," *Environmental Ethics* 30 (2008): pp. 361–79.

Utsler, David, "Paul Ricoeur's Hermeneutics as a Model for Environmental Philosophy," *Philosophy Today* 53 (2009): pp. 173–8.

van Toorn, Willem, *Een leeg landschap* (Amsterdam: Querido, 1988).

van Toorn, Willem, *Leesbaar landschap* (Amsterdam: Querido, 1998).

van Toorn, Willem, *De Rivier* (Amsterdam: Querido, 1999).

Verhoog, Henk, "Lezen in het boek van de natuur," in Wouter Achterberg (ed.), *Natuur: uitbuiting of respect?* (Kampen: Kok Agora, 1989), pp. 141–59.

Chapter 10

Who am I, who are these People, and what is this Place? A Hermeneutic Account of the Self, Others, and Environments

David Utsler

Introduction

Placing—or rather, replacing—ourselves in the natural world is a daunting and complicated task. Stephanie Mills speaks of "going back to nature when nature's all but gone."[1] Human beings are frequently presented as being "out of place" in the natural world. But aren't human beings also a part of nature? How can we speak of humans as outside of nature; or "nature" as if it is something wholly differentiated from the human? Let us define here "replacing" ourselves in the natural world as achieving a symbiotic balance with the other than human world; a balance that fosters flourishing in human beings and the other members of the natural world in which we dwell. Could we say that by replacing ourselves in nature in these terms that we are recreating ourselves and the natural world as well? Are we not restoring ourselves in a balanced relationship with nature and providing a chance to restore nature in the process?

I will address the human relationship to the other than human world by proposing a hermeneutic account of this relationship. Specifically, I would like to focus on Paul Ricoeur's "hermeneutics of the self" as a paradigm from which our relationship to nature can be constructed and explained. I will begin with Ricoeur's account of personal identity and selfhood; from there I will turn to the hermeneutical tension that exists between distance and dwelling (distanciation and belonging) before moving to narrative (environmental) identity; and finally narrative identity will give way to the ethical implications following from the concept of narrative itself.

---

[1] This was the title of Mills' keynote address at the 2007 International Association for Environmental Philosophy. Stephanie Mills, "Going Back to Nature When Nature's All But Gone." This address appears in the IAEP journal, *Environmental Philosophy* 5 (2008): pp. 1–8.

Environmental Identity[2]

In his introduction to Ricoeur's philosophy of translation, Richard Kearney notes that in many of Ricoeur's works he "exploded the pretensions of the cogito to be self-founding and self-knowing. [Ricoeur] insisted that the shortest route from self to self is through the other."[3] Don Ihde says, "For Ricoeur it is impossible" for the self to know itself "directly or introspectively. It is only by a series of detours that [one] learns about the fullness and complexity of [one's] own being and of [one's] relationship to Being."[4] The self as interpreted—the hermeneutic self—is not the autonomous subject, but a complex multi-dimensional self constituted in relation to the *other*. To the extent that relationship to the environment is constitutive of personal identity, there is a component of personal identity that I call "environmental identity." Environmental identity is one's self-understanding, or self-interpretation, in relationship to one's environment.[5] In order to more fully develop and explain a hermeneutic notion of environmental identity, I must first outline some relevant features of Ricoeur's philosophy of selfhood and identity. When brought to bear on the environment, Paul Ricoeur's hermeneutics of the self provides a framework through which one can begin to understand oneself as a vital member of the ecological community and come to the awareness that the ecological/environmental self is one dimension of the multi-dimensional self.[6]

Oneself as Another is Ricoeur's most complete explication of his hermeneutics of the self. He begins *Oneself as Another* with this statement: "By the title, I wish to designate the point of convergence between the three major philosophical intentions that influenced the preparation of the studies that make up this book."[7]

---

[2]   For more on "environmental identity," see my essay "Paul Ricoeur's Hermeneutics as a Model for Environmental Philosophy," *Philosophy Today* 53 (2009): pp. 173–8.

[3]   Richard Kearney, "Introduction: Ricoeur's Philosophy of Translation," in Paul Ricoeur, *On Translation* (New York: Routledge, 2006), p. x.

[4]   Don Ihde, *Hermeneutic Phenomenology: The Philosophy of Paul Ricoeur* (Evanston: Northwestern University Press, 1971), p. 7.

[5]   "Environment" itself can be an ambiguous term. In my conception of environmental identity, "environment" can be broadly construed so as to include any kind of "environs" in which we dwell.

[6]   While my approach is decidedly philosophical, there is a space for a theological component that could be introduced. David C. McDuffie's chapter, "Concern for Creation: A Religious Response to Ecological Conservation," in this volume (Chapter 11) is one such example. He writes: "Religious responses to the natural environment necessarily involve a sense of connection and an orientation of religious practitioners to the natural systems of which they are inextricably a part. This orientation or reorientation can potentially lead to a renewed sense of place and consequently to the restoration of ecosystems as well as to the restoration and maintenance of sustainable human communities on the behalf of divine creation."

[7]   Paul Ricoeur, *Oneself as Another*, trans. Kathleen Blamey (Chicago: The University of Chicago Press, 1992), p. 1.

These three intentions are: (1) "the primacy of reflective meditation over the immediate positing of the subject." Self-understanding is not found immediately in the *cogito*, but in reflexive meditation; (2) to distinguish between *idem* and *ipse* identity and to show the dialectic between the two. The former, Ricoeur says, refers to permanence in time whereas the latter does not imply an "unchanging core of personality"; (3) "the dialectic of the *self* and the *other than self*." This third intention, Ricoeur notes, most closely relates to the title *Oneself as Another*. The constitution of selfhood "implies otherness to such an intimate degree that one cannot be thought of without the other, that instead one passes into the other."[8] So in Ricoeur's philosophy, the self is largely constituted indirectly through the other. This *other*, it should be evident, is *every other* through which we pass that we consider ourselves in relation to and through which we gain some sense of self-understanding. As I will note later, otherness is not limited to human others but may also include non-human others as well, which might be recognized to be individual entities (e.g. a tree or a river) or even more complex ecosystems or landscapes.

Thus, we return to the notion of detours taken through reflective meditation. The fact that we interpret ourselves through detours is revealed in acts of self-disclosure. If someone says "tell me about yourself" or "I would like to get to know you better," you do not respond with an expressionless face, *"cogito ergo sum"* (at least not if you wish to be taken seriously). Instead, the response is self-descriptive—"I am..." followed by where you are from, what you like to do, what you do for a living, who your family is or any other of a number of descriptors by which you identify or recognize yourself both to yourself and others. To the extent that these descriptors are environmental, identity can be called environmental identity. In short, we recognize ourselves to a great extent through others to whom or which we are in relation. But can we speak of a "relation" with an environment or particular members of an environment, such as a tree? If by "relation" we require conscious interchange between beings, then the answer is "no." I understand relation to be that moment when the other being ceases to be a mere object in my "environs" and becomes a meaningful being in my life,[9] something I transcend myself toward—an "other" that I consider a part of who I am. This will become clearer as we explore the third intention below.

The second intention is the dialectic of *idem* and *ipse*, which refers to the constitution of selfhood and identity through the dialectic of the person as the same over time and as changing over time. The primary distinction in the dialectic of *idem* and *ipse* identity is that the former is unitary whereas the latter indicates plurality in the recognition of the self. The person who takes her or his first breath

---

8   *Ibid.*, p. 3.

9   I call to mind Martin Buber's reflection on contemplating a tree whereby he is "drawn into a relation, and the tree ceases to be an It. The power of exclusiveness has seized me." Martin Buber, *I and Thou*, trans. Walter Kaufmann (New York: Touchstone, 1996 [1970]), p. 58.

at the beginning of life is the selfsame person who takes the final breath at the end of life. Yet that person who can be recognized as the same has also experienced a lifetime of change. One's *self* emerges in the dynamic of that dialectic. Consider for a moment the apparent randomness of existence. Where one makes a living, whom one marries, etc. takes place as the result of a convergence of drastically and radically contingent factors. That is, the narrative that makes up our lives is one of many possible worlds that might have been. So imagine that the combination of contingent events and personal choices made by an individual's parents resulted in that person being born and brought up in an entirely different place than she or he really had been. It might have been a different part of the same country or an entirely different place on the globe. Although this person would have been the identical person in terms of *idem* identity (possessing even the exact same DNA), she or he would have been a different person in terms of the question "who?" in relation to her or his environment.

Thus far, environmental identity is primarily anthropocentric in description. I use "anthropocentrism" here to refer to a vantage point rather than an indicator of moral status. Val Plumwood says, "I think it can be conceded that it is impossible for humans to avoid a certain kind of human epistemic locatedness. Human knowledge is inevitably rooted in human experience of the world, and humans experience the world differently from other species."[10] So under the rubric of reflective meditation and analytical detours, an environmental identity is yielded in relationship to and alongside non-human others. The self is understood through a reflective analysis within the life-world. Turning to Ricoeur's third intention in *Oneself as Another*, however, a different vantage point is discovered. If it is true, as I shall elaborate, that one is one's self to the extent that one is also *other*, then it is possible to speak of a kind of ecocentric or biocentric vantage point from which human beings can see within the very concept of personal identity.

The Dialectic of the Self and the *Other than Self*

The dialectic of the self and the other than self (the third intention) is understood as more than a mere comparison of the self with the other. The otherness to which Ricoeur refers is such that it is actually constitutive of selfhood. It is otherness "to such an intimate degree that one cannot be thought of without the other, that instead one passes into the other,"[11] as quoted above, implying that one is "oneself inasmuch as being other."[12] Can one construe the environment as another self or one's other self? Do we interpret ourselves inasmuch as we *are* the other of the natural environment? Ricoeur refers to "the *polysemic* character of otherness"

---

[10]   Val Plumwood, *Environmental Culture: The Ecological Crisis of Reason* (New York: Routledge, 2002), p. 132.

[11]   Ricoeur, *Oneself as Another*, p. 3.

[12]   *Ibid.*, p. 3.

which implies that the "Other [is] not reduced, as is too often taken for granted, to the otherness of another Person."[13] Already in Ricoeur's thought, otherness that is constitutive of personal identity is open to non-human others as well. It seems experientially evident that we do interpret the self through many things besides other persons. Are there not places or things in each of our lives that we cannot think of ourselves without and that we consider as part of who we are and by which we define and identify ourselves?

The experience of self-interpretation through the other of the natural environment can be understood in terms of the body and vital being.[14] This reality is manifested through two seemingly contradictory states of being—a holistic, integral interrelatedness on the one hand and resistance on the other. The contradiction is only apparent at first glance, for integration and resistance correspond to the tension of the same and the other in our relationship to the natural environment. I exist as oneself as another and I exist as oneself distinct from all others. Let us briefly explore these two aspects of bodily and vital being and how they relate to the dialectic of the self and the other than self.

The first aspect can be described by many terms—e.g. holism, interrelatedness, interdependency, integration. What all of these terms suggest is that all life is joined together at the biotic level. Indeed, as to the basic matter of all physical existence, we even share something with non-living things. As human beings, we may be distinguished and individuated by many characteristics, but we belong (are placed, we might say) within a larger biotic community. We are bound to an interdependent relationship to other entities; a relationship marked by shared and mutual vitality. Any amount of rationality or intelligence does not make that any less the case. With even minimal informed reflection, one can sense solidarity and even intimacy with one's environment and recognize that the other of the natural environment is an "other" without which one cannot think of oneself.

In environmental philosophy, such a holistic approach is presented as the basis for an environmental ethic. At this point in the exploration of the self constituted as an environmental identity, I am making no ethical claims resulting from interrelatedness. While I cannot think of myself without the environmental other to the point that I am constituted in my selfhood by being the environmental other (oneself *as* another), no particular ethical demand is yet placed upon me by this fact alone. This will come later through the operation of narrative. Through narrative, one forges the various parts of life into a story with a coherent meaning. In the inevitable "moral of the story," ethical demands are presented to us. Before we can look to narrative, however, there is still more ground to cover.

---

[13]   *Ibid.*, p. 317; italics original.

[14]   I remain indebted to my former professor, John R. White, for numerous conversations over the years concerning vital being. Cf. John R. White, "Lived Body and Ecological Value Cognition," in Suzanne L. Cataldi and William S. Hamrick (eds), *Merleau-Ponty and Environmental Philosophy: Dwelling on the Landscapes of Thought* (Albany: State University of New York Press, 2007), pp. 177–89.

Whereas I exist in a holistic, integrated relationship to the natural environment marked by shared and mutual vitality, I am also a body sharing space and place with other bodies. I experience what Ricoeur calls "resistance" as a "degree of passivity" that attests to my own existence as well as to the existence of the world external to myself. Resistance here should not be understood in a violent sense or any other negative conception. Resistance refers to the mutual dwelling of various bodies in space and place. The experience of resistance in relation to the natural environment manifests as self-recognition through a dialectic of my own inner intimacy and the world in which I dwell so that "the body is revealed to be the mediator between the intimacy of the self and the externality of the world."[15] As I encounter bodily the other of the natural environment precisely as other, my own sense of selfhood and personal identity is heightened in a deeper self-awareness.

Bodily and vital existence through integrated interdependence and resistance provide one mode of understanding the natural environment as the other through which we pass and by which personal identity is constituted. When identity is constituted in intimate relationship to the environment, this component of personal identity we may call environmental identity. Understood most profoundly in Ricoeur's third philosophical intention related to the very title *Oneself as Another*, environmental identity can be defined as the dialectic of the self and the other than self when the other than self is one's environment.

Distance and Dwelling

Conceptualizing the human relationship to the natural environment runs into some difficulties. Philosophical reflection on environments is frequently engaged in attempts to make sense out of human relationships with those environments. We consider ourselves a part of nature, yet we objectify nature by speaking of human contact or interference with it. This reveals a tension between what I will call simply "distance" and "dwelling."[16] In order to speak of nature or reflect upon the actual experience, we necessarily distance ourselves from the pre-reflective experience of dwelling in the life-world. Other environmental philosophers have noted this paradox.[17]

---

[15]   Ricoeur, *Oneself as Another*, p. 322.

[16]   For a somewhat different, but related discussion on the topics of distance and dwelling, see Forrest Clingerman, "The Intimate Distance of Herons: Theological Travels through Nature, Place, and Migration," *Ethics, Place & Environment* 11 (2008): pp. 313–25.

[17]   See Kenneth Maly, "*A Sand County Almanac*: Through Anthropogenic to Ecogenic Thinking," in Bruce V. Foltz and Robert Frodeman (eds), *Rethinking Nature: Essays in Environmental Philosophy* (Bloomington and Indianapolis: Indiana University Press, 2004), pp. 289–301. See also David Abram, *The Spell of the Sensuous: Perception and Language in a More-Than-Human World* (New York: Vintage Books, 1996), pp. 40–1.

This problem has already been considered more generally as a hermeneutic problem. Ricoeur identified it as an "antinomy" at the heart of Gadamer's hermeneutics, which he describes as "the opposition between alienating distanciation and belonging."[18] We appear to be left with an "alienating distanciation" that "destroys the fundamental and primordial relation whereby we belong to and participate in the historical reality that we claim to construct as an object."[19] Ricoeur rejected distanciation and belonging as alternatives between which we must choose or that must necessarily exclude each other. He argued that there is a "productive distanciation" that is not at odds with "participatory belonging."

While we may certainly say that there are no substitutes for the direct experience of nature, it would be incorrect to think that all valuable understanding of nature can only be grasped in this direct manner. What would be the value or purpose, for example, of nature writing? How would we explain, for instance, the impact of books such as Leopold's *A Sand County Almanac*?[20] For Leopold, wouldn't writing about his experience and encounter with the land distance himself from the actual experience thereby alienating himself from it as an author? Beyond the distance of the author from the experience, the text creates a further distance of the text from the author in the reader. Is the reader now even more alienated (distanced) from the direct experience than even the author who tried to capture it by writing? The point is that stepping back from or stepping out of "direct" experience does not render direct experience null. Distanciation serves as a necessary moment of signification in order to deepen our understanding of the direct experience by appropriation—i.e. making it our own in a new way. A fuller meaning or interpretation of nature may only be accessible through the function of distanciation. In other words, the original experience of dwelling, being "placed" in nature, can be further enriched when, following distanciation, we are replaced in nature through hermeneutical appropriation. Our participation in the natural environment, especially with regard to its meaning, must be a participatory belonging through distance. Distance and dwelling comprise a fruitful tension that creates (or recreates) the human relationship to the environment in which humans reside.

Distanciation and belonging are inextricably linked to self-interpretation, thus lending greater dimension to environmental identity. Distanciation is a moment of belonging insofar as it is a movement towards appropriation culminating in self-understanding. Ricoeur writes:

> By "appropriation," I understand this: that the interpretation of a text culminates
> in the self-interpretation of a subject who thenceforth understands himself

---

[18]   Paul Ricoeur, *From Text to Action: Essays in Hermeneutics II*, trans. Kathleen Blamey and John B. Thompson (Evanston: Northwestern University Press, 1991), p. 75.

[19]   *Ibid.*

[20]   Aldo Leopold, *A Sand County Almanac: With Essays on Conservation from Round River* (New York: Ballantine Books, 1966).

> better, understands himself differently, or simply begins to understand himself
> … Here hermeneutics and reflective philosophy are correlative and reciprocal.
> On the one hand, self-understanding passes through the detour of understanding
> the cultural signs in which the self documents and forms itself … In short, in
> hermeneutical reflection—or in reflective hermeneutics—the constitution of the
> *self* is contemporaneous with the constitution of *meaning*.[21]

What Ricoeur says here about the text equally applies to the encounter with environments. Without further explicating it here, I propose that the "constitution of meaning" that follows from the interpretation of environments and our dwelling therein equally results in self-interpretation in which the self is constituted. Such self-interpretation that follows the interpretive encounter with environments constitutes the self as an environmental identity.

Narrative Environmental Identity

Environmental identity, as with all aspects of personal identity, is dynamic and fluid, insofar as the constitution of the self is temporally bound. All of the various aspects that constitute personal identity are related to time, to memory, and future aspirations. Past, present, and future, gathered together through emplotment, constitute the self as a narrative identity where we become "authors and readers of our own lives."[22] Narrative environmental identity is narrative identity as a story told and understood in relation to environments. Narrative sheds further light on the previous problematic of distanciation and belonging. I am placed in the natural world through direct experience. Even in the midst of the experience I am engaged in the process of reflection—i.e. I am experiencing the natural world while simultaneously taking in the meaning of the experience. The immediate experience places me in the natural environment while the reflexive impulse replaces me in it. Continued reflection on my place in the natural environment—every moment, each physical step, each new sensation—is gathered together into a story that weaves a tapestry of meaning making the direct experience of my placement in the natural environment a narrative environmental identity. All of the disparate parts become conjoined into a story that relates me to my environment. Narrative, therefore, is a productive form of distanciation that allows me to be replaced into the experience by means of appropriation, granting me a deeper level of insight into the original experience—appropriation by means of storytelling. My self-understanding is the consequence of the mediation of the narrative constituting a component of selfhood that can be characterized as a narrative environmental identity.

---

[21]   Ricoeur, *From Text to Action*, pp. 118–19; italics original.

[22]   Kearney, "Introduction: Ricoeur's Philosophy of Translation," p. xix.

Narrative and Ethics[23]

Earlier I noted that environmental identity placed no direct ethical demands upon me. Yet, the constitution of selfhood does have ethical import. If I am constituted in relation to the other, then my action toward the other is not ethically irrelevant. For if the constitution of my *self* derives from the other then my action toward the other says something ethically relevant about my *self*. Action reflects back upon the actor and, thus, becomes evaluative. We now have a framework within and around which we can begin to build an ethic. Specific moral norms are always going to be disputable. I do not suggest here that within environmental hermeneutics in general and environmental identity in particular, we have now discovered *the* method which, when correctly applied, will yield all the right ethical and moral answers to environmental concerns. What environmental hermeneutics and environmental identity do provide is a framework for critical discourse. And it is through narrative, in particular, that we are eventually led to ethical prescription.

I will follow a three-fold pattern suggested by Ricoeur of description, narration, and prescription.[24] Description is the phenomenology of events and actions, narration is the hermeneutic configuration of the events and actions into a meaningful story that likewise constitutes identity, and prescription is the ethical dimension of the narrative. Narrative has the function of mediating between phenomenological description and ethical prescription. Narration as the mediating term assists us in avoiding deriving an ethical *ought* directly from a phenomenological *is*. Narrative is not itself ethical prescription nor does it prescribe a normative course of action, rather "narrativity serves a propaedeutic to ethics," because, as Ricoeur points out, "there is no ethically neutral narrative."[25] Narrative has an inherent evaluative quality. For the individual, narrativity shines an ethical light in terms of choices about courses of action one may choose in life as such choices impact the natural world. Hence, one "places" and "replaces" oneself in nature by virtue of the story one tells about it and, thus, also identifies oneself in the story. Narrativity also has a social and political dimension. We share narratives, conduct discourse within and about our shared narratives (as well as our

---

[23]   An article by Galen Strawson appeared in 2004 challenging the notion of narrativity across several disciplines. See Strawson, "Against Narrativity," *Ratio* 17 (2004): pp. 428–52. For Strawson, experiencing or conceiving one's life as a narrative may not be desirable and can be ethically questionable. However he offers no serious engagement with the detail of Ricoeur's arguments on which I am drawing, and hence fails to rebut them. While Strawson's critique calls for a considered response, I cannot fully address Strawson's arguments here.

[24]   Ricoeur, *Oneself as Another*, p. 114. One cannot help but notice the connection between the triad of describing, narrating and prescribing with prefiguration, configuration, and refiguration (or *mimesis₁*, *mimesis₂* and *mimesis₃*) as proposed in Ricoeur's *Time and Narrative, Vol. 1* (Chicago: The University of Chicago Press, 1983).

[25]   *Ibid.*, p. 115.

disparate ones), and organize ourselves as a communal narrative environmental identity. The quest for "the good life with and for others in just institutions"[26] (Ricoeur's definition of the "ethical intention") is an end of the narrative function. Environmental hermeneutics includes the natural world within the horizon of the "with and for others" with which we exist.

As a model for actions that are to be ethically evaluated, we now return to the triad of phenomenological description, hermeneutical narration, and ethical prescription. Phenomenological description corresponds to the notion of the prefigurative character ($mimesis_1$) of events and actions. Narration corresponds to configuration ($mimesis_2$), or emplotment, through which events and actions are "gathered together" into a meaningful story. It is precisely this "gathering together" that lends itself to prescription. Prescription in some sense corresponds to refiguration ($mimesis_3$) in that there is a sense of valuation that takes place. Refiguration is a response to the narrative configuration of a life. As a life is configured into a meaningful story, the response of the refiguration of one's life would seem to be a giving of an "ethical character" to one's life. As Ricoeur says, "the idea of gathering together one's life in the form of a narrative is destined to serve as a basis for the aim of a 'good' life ... How, indeed, could a subject of action give an ethical character to his or her own life taken as a whole, if this life were not gathered together in some way, and how could this occur if not, precisely, in the form of a narrative?"[27]

Narrative environmental identity and the threefold mimesis referred to above may also be employed in terms of issues of concern to environmental justice studies. Environmental ethics concerns not only other than human entities, but human beings that are disproportionately burdened by the costs of environmental devastation. A communal narrative environmental identity—a group's communal understanding of its being in relation to its environment—serves as an indicator for that group's aspirations for what it understands to be the "good life" for itself. The narrative environmental identity forms the foundation for what the group understands to be its duties and obligations to one another in the context of its environment for the attaining or sustaining of what it takes to achieve the good life. What the "good life" is for a particular people is directly related to their narrative. This good life must be aspired to "with and for others," which would indicate a cooperative and collaborative model, especially when competing narratives, desires, and so forth conflict. This suggests perhaps a Habermasian communicative or discourse model. Lastly, then, the foregoing must be embodied in just institutions; institutions being understood as the various social arrangements that make up a democratic society.

---

[26] *Ibid.*, p. 172.

[27] *Ibid.*, p. 158.

Environmental Hermeneutics and Critique

A critical element is introduced naturally into environmental hermeneutics making it also a "critical environmental hermeneutics."[28] There is no single interpretation of the natural environment that must be identified as the only right one, but it is equally the case that not all interpretations are valid. Some may even be unethical or result in unethical behavior, both toward non-human entities as well as other human beings. As Robert Mugerauer has noted, hermeneutics is about "finding the valid criteria for polysemy within the fluid variety of possibilities."[29] Interpretation opens up many worlds and many narratives. Within the finitude of our disparate hermeneutical horizons, we must be able to identify these varieties of possibilities, mediate the conflict of interpretations[30] existing between valid interpretations, and indentify "false consciousness" present in arbitrary or invalid interpretations as well as the criteria for indentifying such interpretations. Incorporating critique into environmental hermeneutics is important for the debates within the realm of environmental philosophy, but would also seem indispensable in the area of environmental policy formation and application. Environmental hermeneutics (including its inherent critical element) is not confined to mere theoretical devices, but has social and political uses. We place and replace ourselves in nature on social levels up to even global levels through environmental law and policy. Hermeneutics and critique is absolutely essential to this process. Critique is another form of distanciation and appropriation. Taking the critical step back, critique becomes "an objective and explanatory segment, in the project of enlarging and restoring communication and self-understanding."[31]

Conclusion

To conclude, through environmental identity we can open up the possibility of creating and recreating ourselves to the end of saving the natural environment from further devastation. Interpreting the self through nature and understanding nature as one's self and other-than-self, I believe to be a crucial necessity for achieving a right relationship to the natural world—to place nature and to place ourselves in nature. And, insofar as we have displaced ourselves or ruptured our relationship to nature, environmental identity provides the tools we need to replace ourselves in the natural world, achieving the symbiotic balance I spoke of at the beginning of

---

[28] Cf. John van Buren, "Critical Environmental Hermeneutics," *Environmental Ethics* 17 (1995): pp. 259–75.

[29] Robert Mugerauer, *Interpreting Environments: Tradition, Deconstruction, Hermeneutics* (Austin: University of Texas Press, 1995), p. xxvii.

[30] Paul Ricoeur, *The Conflict of Interpretations: Essays in Hermeneutics* (Evanston: Northwestern University Press, 1974).

[31] Ricoeur, *From Text to Action*, p. 35.

this chapter. By doing so, we can have a more robust understanding of the self, a just relationship to others, especially with regard to the distribution of environmental burdens in terms of environmental justice, and, finally, a relationship to the Earth that fosters flourishing in humans and non-human nature.

Bibliography

Abram, David, *The Spell of the Sensuous: Perception and Language in a More-Than-Human World* (New York: Vintage Books, 1996).

Buber, Martin, *I and Thou*, trans. Walter Kaufmann (New York: Touchstone, 1996 [1970]).

Clayton, Susan and Susan Opotow, eds, *Identity and the Natural Environment: The Psychological Significance of Nature* (Cambridge, MA: The MIT Press, 2003).

Clingerman, Forrest, "The Intimate Distance of Herons: Theological Travels through Nature, Place, and Migration," *Ethics, Place & Environment* 11 (2008): pp. 313–25.

Cohen, Richard A. and James L. Marsh, eds, *Ricoeur as Another: The Ethics of Subjectivity* (Albany: State University of New York Press, 2002).

Gadamer, Hans-Georg, *Truth and Method*, 2nd revised edn, trans. Joel Weinsheimer and Donald G. Marshall (London and New York: Continuum Press, 1975 [1989, 2004]).

Gadamer, Hans-Georg, *Philosophical Hermeneutics*, trans. and ed. David E. Linge (Los Angeles: University of California Press, 1976).

Habermas, Jürgen, *The Theory of Communicative Action, Vol. 1: Reason and the Rationalization of Society*, trans. Thomas McCarthy (Boston: Beacon Press, 1984).

Habermas, Jürgen, *Between Facts and Norms: Contributions to a Discourse Theory of Law and Democracy*, trans. William Rehg (Cambridge, MA: The MIT Press, 1996).

Ihde, Don, *Hermeneutic Phenomenology: The Philosophy of Paul Ricoeur* (Evanston: Northwestern University Press, 1971).

Ihde, Don, *Expanding Hermeneutics: Visualism in Science* (Evanston: Northwestern University Press, 1998).

Kearney, Richard, "Introduction: Ricoeur's Philosophy of Translation," in Paul Ricoeur, *On Translation* (New York: Routledge, 2006), pp. xvii–xx.

Leopold, Aldo, *A Sand County Almanac: With Essays on Conservation from Round River* (New York: Ballantine Books, 1966).

Maly, Kenneth, "*A Sand County Almanac*: Through Anthropogenic to Ecogenic Thinking," in Bruce V. Foltz and Robert Frodeman (eds), *Rethinking Nature: Essays in Environmental Philosophy* (Bloomington and Indianapolis: Indiana University Press, 2004), pp. 289–301.

Mills, Stephanie, "Going Back to Nature when Nature's All But Gone," *Environmental Philosophy* 5 (2008): pp. 1–8.

Mugerauer, Robert, *Interpreting Environments: Tradition, Deconstruction, Hermeneutics* (Austin: University of Texas Press, 1995).

Plumwood, Val, *Feminism and the Mastery of Nature* (New York: Routledge, 1993).

Plumwood, Val, *Environmental Culture: The Ecological Crisis of Reason* (New York: Routledge, 2002).

Ricoeur, Paul, *The Conflict of Interpretations: Essays in Hermeneutics*, ed. Don Ihde (Evanston: Northwestern University Press, 1974).

Ricoeur, Paul, *Time and Narrative* (3 vols, Chicago: The University of Chicago Press, 1983, 1990).

Ricoeur, Paul, *From Text to Action: Essays in Hermeneutics II*, trans. Kathleen Blamey and John B. Thompson (Evanston: Northwestern University Press, 1991).

Ricoeur, Paul, *Oneself as Another*, trans. Kathleen Blamey (Chicago: The University of Chicago Press, 1992).

Ricoeur, Paul, *Memory, History, Forgetting*, trans. Kathleen Blamey and David Pellauer (Chicago: The University of Chicago Press, 2004).

Ricoeur, Paul, *The Course of Recognition*, trans. David Pellauer (Cambridge, MA: Harvard University Press, 2005).

Strawson, Galen, "Against Narrativity," *Ratio* 17 (2004): pp. 428–52.

Taylor, Charles, *Sources of the Self: The Making of Modern Identity* (Cambridge, MA: Harvard University Press, 1989).

Utsler, David, "Paul Ricoeur's Hermeneutics as a Model for Environmental Philosophy," *Philosophy Today* 53 (2009): pp. 173–8.

Van Buren, John, "Critical Environmental Hermeneutics," *Environmental Ethics* 17 (1995): pp. 259–75.

White, John R., "Lived Body and Ecological Value Cognition," in Suzanne L. Cataldi and William S. Hamrich (eds), *Merleau-Ponty and Environmental Philosophy: Dwelling on the Landscapes of Thought* (Albany: State University of New York Press, 2007), pp. 177–89.

Chapter 11

Concern for Creation: A Religious Response to Ecological Conservation

David C. McDuffie

Introduction

The purpose of this chapter is to examine the Christian concept of "creation" in relation to efforts (with particular emphasis on an American setting) to address ecological conservation from a theistic perspective. In short, "creation" has become a valuable concept for dialogue between religion and natural science for the common goal of protecting and sustaining ecological systems. I develop my argument in two ways. First, I focus on the pragmatic value of the use of the term "creation" in relation to the potential for religion to make a positive contribution toward alleviating current ecological crises. Second, while I do not intend to provide an exhaustive explication of the theological notion of creation, I discuss the theological and ecological implications of the attempt to incorporate, through the use of a particular understanding of Christian creation, ecological concepts into forms of religious expression that address current ecological problems. My argument centers on the assumption that the various uses of "creation" are a potentially valuable means through which ecological issues can be effectively addressed from a religious perspective and that the further theological development of "creation" in an ecological context can potentially lead to an invaluable contribution to alleviating contemporary ecological crises.

The Concept of "Creation" and Conservation

"Creation," as I am using it here, is in no way related to "creationism" and need not perpetuate the perceived conflict between religion and science. However, while "creation" can be informed by a scientific perspective, it does not have a scientific agenda. In short, by "creation," I mean the theological understanding that the world in which we live and upon which we depend for the continuation of life is, in some sense, a gift of God. All living things are contingent creatures owing existence and sustenance to the interaction of the biotic and abiotic components of our planet's ecosystems, the source of which is understood to be the only self existent and necessary "Being." As a theological and metaphysical claim, the understanding of creation as divine gift does not necessarily conflict with the empirical claims of the natural sciences. This is not to say that theological interpretations are not potentially informed by empirical evidence; however, while they are inseparably related in the task of constructing ecological theologies, theology and natural science are distinct disciplinary endeavors.

Of course, it is not necessarily obvious that the application of the concept of "creation" will be conducive to addressing contemporary ecological concerns. For instance, in 1967 Lynn White, Jr. published an article in the journal *Science* entitled "The Historical Roots of Our Ecologic Crisis"[1] in which he claims that the use of this theological concept has been antithetical to addressing contemporary ecological concerns. In brief, White's thesis is that Christianity bears the greatest culpability for the perpetuation of an anthropocentric attitude toward nature, which, when applied to the merging of modern science and technology, accounts for our modern ecological crises. This anthropocentric attitude, he argues, is grounded in the Christian understanding of the Scriptural accounts of creation. Conflating aspects of the two Biblical creation accounts,[2] he offers his understanding of the Christian interpretation of divine creation: "God planned all of this explicitly for man's benefit and rule: no item in the physical creation had any purpose save to serve man's purposes. And, although man's body is made of clay, he is not simply part of nature: he is made in God's image."[3] From this, he concludes that "in its Western form, Christianity is the most anthropocentric religion the world has seen."[4] Therefore, White claims, the Biblical creation stories, as interpreted especially through Christian tradition, are the source of our current ecological problems in the West.

According to White, this anthropocentric influence has permeated the way that Western societies view humankind's relationship with the natural environment. In his words,

> Our science and our technology have grown out of Christian attitudes toward man's relation to nature which are almost universally held not only by Christians and neo-Christians but also by those who fondly regard themselves as post-Christians. Despite Copernicus, all the cosmos rotates around our little globe. Despite Darwin, we are *not*, in our hearts, part of the natural process. We are superior to nature, contemptuous of it, willing to use it for our slightest whim.[5]

White concludes: "More science and more technology are not going to get us out of the present ecologic crisis until we find a new religion, or rethink our old one."[6]

---

[1]   Lynn White, Jr., "The Historical Roots of Our Ecologic Crisis," *Science* 155, no. 3767 (1967): pp. 1203–7.

[2]   Genesis 1–2:3 and Genesis 2:4–2:25.

[3]   White, "The Historical Roots," p. 1205.

[4]   *Ibid.*, p. 1205.

[5]   *Ibid.*, p. 1206.

[6]   *Ibid.*, p. 1206.

I mention White here because his article serves as a watershed moment in the discussion of Christianity and ecology, and his critique of the anthropocentric interpretations of the concept of "creation" still holds strong. Given the contemporary potential for human induced ecological destruction, through the application of science and technology, the theological understanding of divinely sanctioned anthropocentrism and unmitigated dominion has very little to offer toward the alleviation of our current ecological crises. However, it is my contention that the theological concept of creation is emerging as a viable means to address issues of ecological concern from a religious perspective,[7] and in what follows, I will focus on the implication of the incorporation of creation in an ecological context, both for Christian communities and for attempts to address contemporary ecological crises.

For Christianity, problems arise when the Bible is treated as a scientific text that can be potentially in conflict with the knowledge produced from the disciplines of natural science. However, the problem is not exclusively a religious one but is also exacerbated by individuals attempting to discount religion (particularly theistic religion) by making broad philosophical assumptions based on empirical evidence. For example, consider the following statement from evolutionary biologist Richard Dawkins in his book *The God Delusion*: "If this book works as I intend, religious readers who open it will be atheists when they put it down."[8] Examples such as this provide evidence that fundamentalism is a problem working on both sides of this unnecessary polarizing debate to determine the source of the natural or created order.

Still, it cannot be denied that the science which grounds the discipline of ecology and, consequently, conservation efforts has been a stumbling block for some conservative Christian communities. Furthermore, environmentalism has often been stigmatized by many of these communities as being at odds with a Christian message. In the words of Richard Cizik, a prominent evangelical leader who has been at the forefront of this conversation, "Environmentalism carries with it baggage—baggage that … it's about big government, kooky religions … all the left-wing ideas that are known to mankind … And evangelicals have said, 'Oh no,

---

[7]   In an aspect of his article that is often underemphasized, White recognizes this potential and suggests that the religious framework of Christian tradition, while contributing to the root of the problem, may also serve as a source for the solution. The quotation corresponding to the previous footnote is clear evidence of this. To reiterate, he writes: "What we do about ecology depends on our ideas of the man-nature relationship. More science and more technology are not going to get us out of the present ecologic crisis until we find a new religion, or rethink our old one" (p. 1206). Elsewhere in the essay, he adds: "Since the roots of our trouble are so largely religious, the remedy must also be essentially religious, whether we call it that or not" (p. 1207).

[8]   Richard Dawkins, *The God Delusion* (New York: Houghton Mifflin Company, 2006), p. 5.

we can't do that. Not us'."[9] In other words, environmentalism has been interpreted as being associated with an ideology that is understood to be counterproductive to Christian belief and worship. Interestingly, "creation" as opposed to "creationism" has emerged as a mediating term, as a euphemism for "environmentalism," in cases of conflict between Christianity and a scientific perspective perceived to be antithetical to Christian religion. For example, Cizik advocates "Creation Care" as opposed to environmentalism when referring to human action on behalf of environmental concern. As a result, by rooting a response to ecological problems in Biblical tradition, advocates of creation care are avoiding the "baggage" associated with environmentalism, which has, unfortunately, caused many Christians to remain at the periphery of this conversation, while remaining open to being informed by scientific evidence.[10]

The resonance and potential effectiveness of this term has also been recognized by individuals approaching issues of conservation from a non-religious perspective. For instance, E. O. Wilson published a book in 2006 entitled *The Creation: An Appeal to Save Life on Earth*, which was written as a letter to a hypothetical Christian minister. Describing Creation as "living Nature," he writes that "it will be necessary to find common ground on which the powerful forces of religion and science can be joined" if the current threats to species on our planet are to be alleviated.[11] The significance of such developments for ecological conservation, which is necessarily an interdisciplinary endeavor, is significant. While it would be naïve to claim that all situations of conflict have been avoided concerning the relationship between Christianity and the science which necessarily informs ecological conservation, "creation," used as a mediating term, has emerged as a remedy for and not a symptom of the problems associated with Christianity's relationship to the natural environment.

While this is a positive development, interpreting nature as creation must move beyond functioning simply as a pragmatic means for dialogue between religion and science if it is to have lasting significance for ecological conservation.

---

[9]   *Religion & Ethics Newsweekly: An Online Companion to the Weekly Television News Program*, "Cover Story: Evangelicals and the Environment" (January 13, 2006, Episode no. 920), www.pbs.org/wnet/religionandethics/week920/cover.html (accessed March 15, 2008).

[10]   In February, 2006, a group of over 85 evangelical leaders formed *The Evangelical Climate Initiative* (see http://christiansandclimate.org). In a statement entitled "Climate Change: An Evangelical Call to Action," these Christian leaders appealed to their religious traditions as well as to the best available scientific evidence in order to address problems associated with human induced global climate change. For a full text of the statement, see http://christiansandclimate.org/learn/call-to-action/. Another excellent example of this type of implementation of the term creation is *The Evangelical Environmental Network and Creation Care Magazine* (see www.creationcare.org/).

[11]   Edward O. Wilson, *The Creation: An Appeal to Save Life on Earth* (New York: W. W. Norton & Company, 2006), p. 165.

Essentially, if the use of the theological concept of "creation" is compatible with contemporary ecological science, it is likely that it will provide an effective means through which to address the human relationship with the ecosystems of which we are inextricably a part. Furthermore, many Christian communities have little or no problem accepting the tenets of modern science including evolutionary theory; therefore, creation used solely as a means to avoid perceived conflicts between religion and ecological science will have a limited impact on conservation efforts from a Christian perspective.

"Creation" in an Ecological Context

What does it mean to speak of a theological concept of "creation" in a contemporary conversation concerned with protecting our natural environment? If it is to have a significant effect, a Christian contribution to ecological conservation must be simultaneously Christian and ecological. On the one hand, any theological use of the Christian concept of "creation" must be in continuity with the tradition of which it is inextricably a part; moreover, contemporary religious expression on behalf of ecological concern must be informed by the best available ecological science.[12] It is necessary to examine the distinctive contribution from Christianity, and religion more generally, in an ecological context and not attempt to shoehorn ecology into forms of religious expression in order to take advantage of organized communities for the purpose of affecting a positive ecological influence. In brief, if the concept of "creation" is to have an effective influence for conservation, the ecological and theological, while distinct, must be understood as inseparably related to each other.

However, given the aforementioned problems concerning Christianity and the natural environment, it may seem as if the necessity of including past tradition into a contemporary Christian expression on ecological matters is a hindrance. Certainly, any contemporary understanding of creation must take into account criticisms such as White's and recognize that there are stages in Christian history where portrayals of theistic creation were interpreted, or rather misinterpreted, in a manner that lends credence to White's account of Christianity as ecologically destructive. However, as many concerned with Christian tradition have noted, White's argument denigrating the environmental applicability of Christian tradition is overstated. As Sallie McFague points out, the legitimacy of arguments such as White's is more accurate in reference to the post-Enlightenment "turn to the self" and the consequent denial of purpose in the natural order than it is in relation to a blanket condemnation of the majority of Western Christian tradition. She writes: "From the earliest days of Christianity, the cosmological context was

---

[12]   Schubert Ogden claims that all Christian theology must be both *appropriate* in relation to previous Christian tradition and *credible* in relation to contemporary experience. See *On Theology* (San Francisco: Harper and Row Publishers, 1986), pp. 4–5.

a major interpretive category along with the psychological and the political. The renewal of creation, the salvation of the individual, and the liberation of the people were all seen as necessary components of the work of God in Christ."[13] In brief, while the divinity of nature is rejected by Biblical and much of the post-Biblical Christian tradition, the cosmos was not devoid of meaning and purpose as it was understood to be sacralized as a creation of God.

Nevertheless, when speaking of the relation of a theological notion of creation to ecological concern, we must be very clear concerning how the religious tradition relates to these contemporary debates. For instance, we must not make the mistake of anachronistically attributing knowledge of ecological science to the writers of the Bible and most of post-Biblical religious tradition. When we return to past tradition in order to recover ecologically credible themes, we are not looking for claims of science but for a theological interpretation of the relationship between an understanding of God and the natural world, which is understood to be God's creation. In this sense, religion provides a means through which to attribute value to the natural or created order. According to Ellen Davis in an introductory essay to the *Green Bible*, "the Bible's own perspective is both material *and* spiritual, for God is the creator of everything that is. Therefore the environmental crisis must be seen as a profound religious crisis; it is a massive disordering of our relationship with the God who created heaven and earth."[14] In fact, using the term "creation" broadens the association of meaning and value with natural environments to include an understanding of ecological communities in relationship with the divine. Such a theological understanding could profitably add a theological dimension to what David Utsler has called our "environmental identity."[15] In essence, then, the theological concept of "creation", as it is conveyed in the Biblical text, affirms the goodness of the non-human natural environment and contributes to the applicability of this aspect of Christian tradition to issues germane to ecological conservation efforts.

Similar arguments can be made with reference to the post-Biblical theological tradition. In *City of God*, Augustine claims "that all things which do not belong to God's own being, though inferior to God, are nevertheless good, and the creation of God's goodness."[16] Similarly, Thomas Aquinas writes: "Every being that is not God, is God's creature. Now *every creature of God is good (I Tim. Iv. 4)*: and God

---

[13]   Sallie McFague, "An Ecological Christology: Does Christianity Have It?" in *Christianity and Ecology*, edited by Dieter T. Hessel and Rosemary Radford Ruether (Cambridge, MA: Harvard University Press, 2000), p. 29.

[14]   Ellen F. Davis, "Knowing Our Place on Earth: Learning Environmental Responsibility from the Old Testament," in *The Green Bible NRSV* (New York: HarperCollins, 2008), pp. 1-58.

[15]   Cf. Chapter 10, this volume.

[16]   Augustine of Hippo, *City of God*, trans. Henry Bettenson (London: Penguin Books, 1984), p. 455.

is the greatest good. Therefore every being is good."[17] In these examples from two Western theological giants, the goodness of the created order is affirmed through the Creator, the Christian theistic understanding of God. As a result, the natural world is endowed with goodness and is not to be degraded as it is a gift of God. This notion of goodness and value can be applied to a contemporary Christian approach to ecological concern; however, it must be qualified with the acceptance that aspects of the tradition could be selected to offer a counter interpretation. Individuals such as Augustine and Aquinas did not approach Christian theology from the perspective of the modern discipline of ecological science, and we would be wrong to hold them to such a standard.

"Creation" as a Theological and Ecological Concept

These examples do not deny the limitations that still persist in relation to applying past Christian tradition to contemporary ecological debates. Because the representatives of much of Christian tradition did not have our understanding of ecological science, we must recognize that they do not always avoid the anthropocentrism that is antithetical to an ecological worldview. As a result, while the traditional theological understanding of the meaning and purpose contained within the created order is integral to a contemporary Christian approach to ecological issues, the theological notion of "creation" must be brought to bear on our current situation. In brief, we must ask what it means to speak of creation in a contemporary ecological context.

Unfortunately, in our contemporary context, there is often the implicit understanding that theistic "creation" is a static concept in that it is an event in the past. In other words, there was a temporally *original* creation yet there is little or no understanding of a *perpetual* or *ongoing* creation. This need not be the case. According to Norman Wirzba, "our penchant to read this doctrine primarily in terms of origins" has resulted in our repeated lack of appreciation for "the nature and character of responsible human life within the world."[18] Furthermore, he writes: "Were the creation a static or completed act, it would make little sense for God to be in relation to it as its inspiration, sustenance, and end."[19] Essentially, the creation can be understood to be continually infused with meaning, significance,

---

[17]   Thomas Aquinas, *Summa Theologica, Part I* in *Basic Writings of Saint Thomas Aquinas: Volume One*, ed. Anton C. Pegis (Indianapolis: Hackett Publishing Company, 1997), p. 45.

[18]   Norman Wirzba, *The Paradise of God: Renewing Religion in an Ecological Age* (New York: Oxford University Press, 2003), p. viii.

[19]   *Ibid.*, p. 25.

and purpose by God, that which is understood by Christian theology to be the source, center, and end of all things.[20]

I am not arguing for an attempt to reinstate Aristotelian notions of formal and final causality in the sense that there is a discernible teleological end toward which the biotic and abiotic components of our ecological systems are tending. However, what is at work here is an effort to move beyond a reductionistic view of our natural environment based solely on a mechanistic understanding of efficient and material causation. Purpose and meaning are granted to ecological entities through an understanding of the theological concept of creation, and an appreciation and awareness of the interconnected aspects of our earthly lives are cultivated through an increased understanding of the scientific discipline of ecology.

For example, I work as a member of the Environmental Stewardship Committee at the Chapel of the Cross Episcopal Church in Chapel Hill, NC. Currently, we are in the process of incorporating what will be our second "Creation Cycle" into the liturgical calendar at the local, parish level. Last year (2009), we completed our first Creation Cycle which consisted of a period of four weeks in the fall explicitly focused on caring for the natural environment, the creation, from a religious perspective.[21] In such a context, awareness of our necessarily ecological context is conveyed through the life of the church and, through this process, individuals of all ages can potentially cultivate concern for natural environments through an orientation toward the theological concept of creation.[22] From this perspective, creation and ecology are intertwined in a common understanding of reverence for the ecological communities of our planet. This sentiment is expressed clearly in a statement from The Episcopal Ecological Network (EpEN), which is the representative organization for addressing issues of concern for the natural environmental from the perspective of the Episcopal Church (USA), describing

---

[20]   I am borrowing language referring to God as the "source, center, and end" of all that exists from William L. Power with whom I studied at The University of Georgia from 2004 to 2008.

[21]   At the 193rd Annual Convention of the Episcopal Diocese of North Carolina in 2009, it was resolved "that the Diocese and its congregations be encouraged to celebrate a liturgical creation cycle of at least four weeks in length at an appropriate time during the liturgical year for the purposes of affirming the sacredness of God's Creation, of spreading hope about God's reconciling work in creation and about an understanding of environmental stewardship and of ecological justice." See http://images.acswebnetworks.com/1/507/ResolutionsonEnvironmentalIssues20002010.pdf.

[22]   Some of the highlights of the Chapel of the Cross's 2009 Creation Cycle included creation themed sermons from members of the clergy, priestly stoles decorated by the children with drawings of the natural environment, an adult education series featuring guest speakers on the relationship between religion and creation/environment, and a Blessing of the Animals Service on the Feast of Saint Francis.

ecology "as the relationship between living beings, the environment in which they live, and God, the Creator."[23]

This should not be read as an attempt at a conflation of the theological concept of "creation" with a scientific understanding of ecology. Instead, in a contemporary, ecological context, "creation," as I am using it here, is a term that is distinct yet inseparably related to the ecological science by which it is informed. While remaining distinct, theological "creation" incorporates ecology into a religious understanding of the proper relationship between humanity and the ecosystems of which we are a part. Consequently, through a proper understanding of creation, the goals of religion and ecological science can be considered as mutually contributing to the common goal of ecological conservation. "Creation" addresses Christian tradition and contemporary science in the form of a theological perspective predicated upon concern for ecological sustainability and informed by the best available ecological science. The result is an approach to ecological conservation that is both uniquely Christian and ecologically credible.

Conclusion

In discussing the Christian concept of "creation," I have attempted to offer a brief introduction concerning the potential for this religious concept to make a valuable contribution to current ecological conservation efforts. In doing so, I have made a distinction between the empirical claims of ecological science and the theological concept of "creation" in order to exhibit the unique ways in which concern for "creation" continues to emerge as a form of religious expression which is informed by the best available ecological science. In addition, this chapter also addresses the themes of "place" and "restoration" discussed in the present book. Religious responses to the natural environment necessarily involve a sense of connection and an orientation of religious practitioners to the natural systems of which they are inextricably a part. This orientation or reorientation can potentially lead to a renewed sense of place and consequently to the restoration of ecosystems as well as to the restoration and maintenance of sustainable human communities on the behalf of divine "creation". Furthermore, the theological understanding of "creation" represented here offers an opportunity to restore meaning and purpose in the natural world, an understanding which has been absent from much of Western science and religion following the Enlightenment.

I have written here from a perspective of the Christian tradition; however, it should be understood that this has not been done from an attempt at exclusivity. While efforts concerned with ecological conservation are necessarily interdisciplinary, any religious response to ecological concern must also be open

---

[23] *The Episcopal Ecological Network*, "Caring for God's Creation: Called to be Stewards," *The Episcopal Ecological Network*, www.eenonline.org/ (accessed December 8, 2008).

to the potential contributions from various religious traditions. I have attempted here to describe one such contribution from a particular religious tradition and it is my contention that religious responses can potentially have an increasingly integral contribution to alleviating ecological degradation.

The realization that conservation calls for more than scientific evidence in order to be successful has led many contemporary scholars of religion to explicate the potential of unique religious contributions to the alleviation of environment problems. For example, Roger Gottlieb ponders whether religion is perhaps a more efficient method through which to change the paradigm of conspicuous consumption that threatens sustainability in societies such as our own in the United States. He writes:

> Secular environmentalists who critique consumerism often (sadly) come off like shrill spoilsports. "Don't," after all, is not much of a basis for a political movement. Religious environmentalists, on the other hand, can offer satisfactions that don't play into the addictive tendencies of always wanting more. The delights of a quiet Sabbath, the peace of a long-term practice of meditation, the joys of celebrating creation in a community of people you know—these cannot be bought or sold, but surely promise more real satisfaction than another trip to the mall.[24]

While ecological science informs us concerning our relationship with the ecological systems upon which we necessarily depend for life, it tells us little if anything about the meaning and significance of the interconnectedness of our lives with the biotic and abiotic entities of our local and planetary environments. On the other hand, one of the goals of religion is to provide commentary on the meaning and significance of the lives of religious practitioners and the world in which these lives are lived. Perhaps more than anything else, the recognition and acceptance of our status as interconnected creatures[25] will (re)orient us to a respect for and desire to better understand our relationship with all of the created, natural order.

Bibliography

Aquinas, Thomas, *Summa Theologica, Part I* in *Basic Writings of Saint Thomas Aquinas: Volume One*, edited by Anton C. Pegis (Indianapolis: Hackett Publishing Company, 1997).

Augustine of Hippo, *City of God*, trans. Henry Bettenson (London: Penguin Books, 1984).

---

[24]   Roger Gottlieb, "The Beginning of a Beautiful Friendship: Religion and Environmentalism," *Reflections* 94 (2007): p. 12.

[25]   See Rowan Williams, "On Being Creatures," in *On Christian Theology* (Malden, MA: Blackwell Publishers, 2000), pp. 63–78.

Davis, Ellen F., "Knowing Our Place on Earth: Learning Environmental Responsibility from the Old Testament," in *The Green Bible NRSV* (New York: HarperCollins Publishers, 2008).

Dawkins, Richard, *The God Delusion* (New York: Houghton Mifflin Company, 2006).

Episcopal Ecological Network, "Caring for God's Creation: Called to be Stewards," *The Episcopal Ecological Network*, www.eenonline.org/.

Gottlieb, Roger, "The Beginning of a Beautiful Friendship: Religion and Environmentalism," *Reflections* 94 (2007): pp. 10–14.

McFague, Sallie, "An Ecological Christology: Does Christianity Have It?" in *Christianity and Ecology: Seeking the Well-Being of Earth and Humans*, edited by Dieter T. Hessel and Rosemary Radford Ruether (Cambridge, MA: Harvard University Press, 2000), pp. 29–45.

Ogden, Schubert, *On Theology* (San Francisco: Harper and Row Publishers, 1986).

Religion & Ethics Newsweekly: An Online Companion to the Weekly Television News Program, "Cover Story: Evangelicals and the Environment," Episode no. 920 (January 13, 2006), www.pbs.org/wnet/religionandethics/week920/cover.html.

White, Jr., Lynn, "The Historical Roots of Our Ecologic Crisis," *Science* 155, no. 3767 (March 10, 1967): pp. 1203–7.

Williams, Rowan, *On Christian Theology* (Malden, MA: Blackwell Publishers, 2000).

Wilson, Edward O., *The Creation: An Appeal to Save Life on Earth* (New York: W. W. Norton & Company, 2006).

Wirzba, Norman, *The Paradise of God: Renewing Religion in an Ecological Age* (New York: Oxford University Press, 2003).

Chapter 12

Recreate, Relate, Decenter: Environmental Ethics and Domestic Animals[1]

Anna L. Peterson

Introduction

In this chapter, I address an issue that does not attract a lot of attention in environmental ethics these days: domestic non-human animals. Many people who write about environmental ethics or animal welfare assume either that the two fields have separate concerns or, more strongly, that their priorities are divergent, perhaps even opposed. Thus a real division has emerged between philosophical thinking about non-human animals, especially domestic ones, and about the natural world more broadly. This division probably cannot be bridged entirely, and in fact it may not need to be. However, we can learn a great deal from looking at the points of connection or at least of mutual illumination between the two fields.

This is not just an interesting task but an important one, because—and this is my argument—attention to domestic animals can help clarify and even recreate our ethical response to non-human nature in (at least) three interconnected ways. First, as we recreate or play with domestic animals, especially companion species, we recreate our attitudes and practices regarding the more-than-human world. I will spend most time on this aspect, which for me establishes the foundation for thinking about other issues. Second, when we relate to domestic animals as partners in play and in work, we are relating to them also as moral agents, and in so doing we recreate not only cross-species relationships but larger communities. Finally, when we decenter ourselves through mutually transformative interactions with animals and reflections on those encounters, we recreate human self-understanding along lines that are both more humble and more ecologically viable.

Before continuing, I should specify that I am not talking about all domestic animals. Non-human species have been domesticated for three main reasons: to supply food and clothing, to do specific kinds of work, and to provide companionship. By far the largest proportion of domesticated animals fall into

[1] I would like to thank Dan Spencer and Bill Jordan for their helpful comments and suggestions on this chapter, and Mark Dixon and Forrest Clingerman for organizing the conference at which it was presented. A few passages in this chapter are taken from Chapter 4 of my book *Everyday Ethics and Social Change: The Education of Desire* (New York: Columbia University Press, 2009).

the first category. However, I am primarily interested in the latter two: work and companionship. And then among all the different species that provide work and companionship, I focus on two: dogs and horses. These are a relatively small subset of all the domestic animals in the world, but they loom very large in terms of their importance to human culture, history, and evolution. Most important, we interact with these species in very special ways, which shed light on the issues I want to discuss here. I am bracketing for now the question of whether this discussion might have any relevance to other domestic animals, especially those that are kept by humans primarily for food. I suspect that our relationships to domestic animals, and their places both in human societies and in larger ecosystems, are too varied to be captured by a single argument or theory. I recognize, however, that any extended discussion of the place of domestic animals in environmental ethics will have to deal with a wider range of domestic animals than I address here.

Environmental Ethics and Animals

The relations and especially the conflicts between environmental ethics and animal ethics are important background for this topic. I will outline very briefly the two primary positions. Philosophers concerned with animal ethics (or welfare, or liberation, or rights) focus on the moral value (or interests, or needs, or rights) of individual animals, sometimes limited to a particular subset defined by characteristics such as sentience or capacity for suffering. Environmental ethicists, on the other hand, are concerned with the moral valuation of non-human nature as a whole, often described by collective nouns such as "ecosystems," "biotic communities," or "the land." Thus while both animal and environmental ethics are concerned with the moral status of (and appropriate attitudes and behavior toward) non-human nature, the former is individualistic in orientation while the latter tends to be holistic. There are, of course, many exceptions to and variations on this general rule. However, overall it holds and, more important, this distinction between individual and holistic approaches is the key to the conflicts between animal and environmental ethics, as they have played out in the scholarly literature and often in practical, political arenas as well.

Environmental ethics does value non-human animals, but it does so not for individual qualities such as sentience but rather on the basis of collective characteristics such as scarcity, ecological significance, or contributions to larger goods such as biodiversity or land health. Thus members of an endangered species or one that is crucial to an ecosystem are more valuable than individuals of a more common or peripheral species. Individual characteristics such as intelligence, sentience, or complexity should not play major roles in moral, political, or scientific judgments. There is a general (though not universal) consensus among philosophers that the "rights and interests of individuals are not a helpful basis for an environmental ethic," because, as Bryan Norton summarizes, the interests of individual organisms bear only a contingent relationship to the primary value of "the healthy functioning and integrity of the ecosystem."[2] In other words, what is good for individuals is not necessarily good for the whole, and the most important criterion for an environmental ethic is the good of the whole. This approach is common to an otherwise diverse set of environmental philosophers, including moral pluralists like Norton, "monists" like Baird Callicott, and a host of other important thinkers including Holmes Rolston, Arne Naess, and many others.

The holistic emphasis within environmental ethics makes the status of domestic animals problematic, at best. Wild animals are morally valuable if and when they contribute to "the healthy functioning and integrity of the ecosystem," not as individuals. Domestic animals, who almost by definition cannot contribute to a healthy ecosystem, thus would appear to lack any moral value whatsoever.[3] In fact, since domestic animals often harm wild ecosystems, an environmental ethic probably should aim to reduce or eliminate them, as when domestic or feral animals threaten rare native plants or degrade entire ecosystems.

There are, however, exceptions, cases in which domestic animals contribute to holistic ecological considerations. Here are a few examples, all involving dogs. The first case is the use by ranchers of livestock guardian dogs to avoid the use of lethal methods such as traps or poison to control wolves, coyotes, cougars, and other predators. This example is ambiguous, since the guardian dogs are protecting domestic animals who probably should not be grazing in the predators' habitat anyway. The dogs' job is less to make a positive ecological contribution than to minimize the damage done. The second example is the use of dogs specially trained to identify particular plant or animal species. Some of these dogs are used to search wild habitats, vehicles, boats, air cargo, and mail for invasive exotic species, including bees in Australia, quagga mussels in California, and Burmese pythons in South Florida.[4] Sniffer dogs also work in airports to find endangered animals transported by smugglers, both live and as "bushmeat."[5] And still other dogs have been trained to identify the scat of particular wild animals, enabling rangers and scientists to locate, monitor, and protect endangered species in Brazil and elsewhere.[6] These cases are fascinating in and of themselves and also as

---

[2]   Bryan G. Norton, "Environmental Ethics and Nonhuman Rights," in *The Animal Rights/Environmental Ethics Debate: The Environmental Perspective*, ed. Eugene C. Hargrove (Albany: SUNY Press, 1992), p. 86 and p. 85.

[3]   I refer to non-human animals as "who," rather than the more common "that" or "which," since "who" emphasizes the agency of such animals—a theme that is central to this chapter.

[4]   http://yosemite.epa.gov/r10/ECOCOMM.NSF/Invasive+Species/Document-Library/$FILE/IS-news18.pdf; http://conservationreport.com/2008/02/26/invasive-species-burmese-pythons-an-invasive-species-in-south-florida-could-spread-to-one-third-of-united-states/.

[5]   http://news.gov.hk/en/category/lawandorder/080114/html/080114en08004.htm.

[6]   www.conservation.org/FMG/Articles/Pages/sniffing_dog_saving_species.aspx.

evidence of the extraordinary working partnerships that dogs and humans form with each other for a huge range of purposes. I will return to this topic later.

Sniffer dogs aside, domestic animals are not morally considerable in most environmental ethics. The individual qualities that animal ethicists value, such as sentience, reason, or emotional complexity, are rarely, if ever, important ecologically. Instead, many environmental philosophers argue, these qualities are valued because of human preferences and priorities. We appreciate in other creatures either traits we value in ourselves, such as intelligence; traits that enable them to bond with us, such as loyalty; or traits that inspire compassion, such as the capacity for suffering. (While wild animals also possess such traits, these are not the source of their ecological or moral value in most environmental philosophies.) Again, the differences between environmental ethics and animal ethics have played out as a conflict between a holistic perspective, on the one hand, and an individualistic perspective, on the other hand.

Searching for Common Ground

In the midst of this sometimes heated debate between animal welfare and environmental ethics, some thinkers have sought common ground. One of the best known, at least within environmental philosophy, is Paul Taylor's ethic based on a "respect for life" that can encompass both individuals and species.[7] Taylor's ethic proposes that "it is the good (well-being, welfare) of individual organisms, considered as entities having inherent worth, that determines our moral relations with the Earth's wild communities of life." Taylor's model is biocentric, because value is found not in ecological wholes but in life itself. Because life is a property only of individual organisms, "we have prima facie moral obligations that are owed to wild plants and animals themselves as members of the Earth's biotic community. We are morally bound (other things being equal) to protect or promote their good for *their* sake."[8] This good is an individual good, inherent in every organism that has a good of its own, which is the case for all animals and plants, according to Taylor. Taylor's primary goal is to undermine anthropocentric ethics that automatically privilege humanness. However, in arguing for an environmental ethic based on the moral value of individuals rather than ecological wholes, he points to an environmental ethic that can value both domestic and wild animals and plants.

---

[7]   Taylor talks of this ethic as both a "respect for life," meaning "life" in a more foundational sense, and as a "respect for nature" more generally. Paul Taylor, *Respect for Nature: A Theory of Environmental Ethics* (Princeton: Princeton University Press, 1986). See also Paul Taylor, "The Ethics of Respect for Nature," in *The Animal Rights/Environmental Ethics Debate*, ed. Eugene C. Hargrove (Albany: SUNY Press, 1992), pp. 95–120.

[8]   Taylor, "The Ethics of Respect for Nature," pp. 95–6.

I find Taylor's argument about inherent worth more plausible as a starting point than a finished ethic. Once we agree that all beings, human and not, have inherent value, we still have a great deal of work to do. What does it mean to respect nature in concrete situations? What kind of practices support and are supported by an ethic of respect for nature? Taylor's overly abstract ethic leaves these questions unanswered, although he does offer one provocative suggestion. The way that we come to know the inherent value of individual organisms, he writes, is by striving to understand their "point of view" and ultimately to look at the world from their perspective.[9] He makes this call to understand non-human "points of view" almost in passing, but in so doing he opens the possibility of a deeper, mutually transformative understanding and interaction. Such interactions are most likely to occur with domestic animals, for most of us. By paying attention to such interactions, environmental ethicists could learn more not only about the inherent value of non-human nature but also about the ways we can and should live with that nature. Taylor somewhat casually drops a ball that I will pick up and run with later.

Several other philosophers have tried more systematically to find common ground between animal and environmental ethics. John Fisher argues that we can value different things, including both individual animals and natural ecosystems, for the same underlying reason. Natural sympathy, according to Fisher, people feel for both animals and natural places, undergirds both environmentalist and animal welfare concerns. "Sympathy for animals, wild as well as domesticated," in fact, plays an important role in many people's concern for the environment broadly conceived, and thus can provide "a partial basis for environmental ethics."[10] Fisher acknowledges that sympathy can be unreliable and biased, far from an infallible guide to correct ethical attitudes or action. It is, however, a powerful enough basis that we should not dismiss sympathy as irrelevant to ethics.

Taylor and Fisher both try to find an overarching moral framework that can value both individuals and groups. Another way to bring animal and environmental ethics together is to acknowledge that perhaps we cannot value very different entities for the same reasons—and that might be okay. Some ecofeminists have made this point, challenging the unproductive dualism of both holists and individualists, neither of which, as Marti Kheel notes, "can see that moral worth can exist *both* in the individual parts of nature *and* in the whole of which they are a part."[11] Similarly, Mary Anne Warren proposes that we combine the animal liberationist concern for sentient individual creatures and the environmentalist concern for ecological wholes. "It is not necessary to choose between regarding

---

[9]   *Ibid.*, pp. 109–10.

[10]   John A. Fisher, "Taking Sympathy Seriously: A Defense of Our Moral Psychology Toward Animals," in *The Animal Rights/Environmental Ethics Debate*, ed. Eugene C. Hargrove (Albany: SUNY Press, 1992), p. 229 and p. 236.

[11]   Marti Kheel, "The Liberation of Nature: A Circular Affair," *Environmental Ethics* 7 (Summer 1985): p. 140.

biological communities as unified systems, analogous to organisms, and regarding them as containing many individual sentient creatures, each with its own separate needs and interests," Warren explains, "for it is clearly both of these things at once."[12]

Warren's expansive view of value echoes the work of Mary Midgley, who has argued persuasively that we can value different entities—individual creatures, species, or entire ecosystems—on different bases, without necessarily falling into serious moral or intellectual contradiction. We should think of moral claims of different types as an overlapping web, Midgley proposes, rather than as concentric circles with an inner core, of which other values are mere extensions.[13] Midgley is mostly concerned with sorting out how we value non-human animals in relation to human beings, but her arguments also elucidate efforts to value non-human animals as individuals in relation to ecological wholes. She acknowledges that differences exist between concern for individual beings and ecological concerns. The real problem here, she notes, may not be how to achieve philosophical consistency but rather the more practical question of how widely we should direct our concern. Should we minimize it by applying it to the smallest number of beings possible? Such an approach is implicit in the arguments of both individualists and holists. In contrast, Midgley contends that moral concern and compassion do not need "to be treated hydraulically ... as a rare and irreplaceable fluid, usable only for exceptionally impressive cases." Rather, they grow with use. Thus, "effective users do not economize on them."[14]

Midgley's work provides a crucial resource for Baird Callicott in his efforts to resolve the conflict he helped ignite.[15] In an essay titled "Animal Liberation and Environmental Ethics: Back Together Again," Callicott regrets the "acrimonious estrangement" between advocates of animal welfare and environmental ethicists that his earlier work helped encourage. The two camps have overlapping concerns, he notes, and "it would be far wiser to make common cause against a common enemy – the destructive forces at work ravaging the nonhuman world – than to continue squabbling among ourselves."[16] While this is a practical concern, Callicott shies away from a purely pragmatist solution, because only a

---

[12]   Mary Anne Warren, "The Rights of the Nonhuman World," in *The Animal Rights/ Environmental Ethics Debate*, ed. Eugene C. Hargrove (Albany: SUNY Press, 1992), p. 206.

[13]   Mary Midgley, *Animals and Why They Matter* (Athens, GA: University of Georgia Press, 1983), pp. 29–30.

[14]   *Ibid.*, p. 31 and p. 144.

[15]   Callicott's most direct contribution to this conflict was his article "Animal Liberation: A Triangular Affair," first published in the journal *Environmental Ethics* in 1980 and later reprinted in Hargrove's edited volume.

[16]   J. Baird Callicott, "Animal Liberation and Environmental Ethics: Back Together Again," in *The Animal Rights/Environmental Ethics Debate* (Albany: SUNY Press, 1992), p. 249.

theoretically coherent ethic provides ground for deciding among competing moral claims. He seeks "a moral theory that embraces both programs *and* that provides a framework for the adjudication of the very real conflicts between human welfare, animal welfare, and ecological integrity."[17] He finds the grounds for such a theory in Midgley's work, especially her concept of the mixed community, which he finds compatible with the holistic land ethic of Aldo Leopold. Drawing on Midgley, Callicott argues that "we are members of nested communities each of which has a different structure and therefore different moral requirements."[18] This "biosocial" theory allows for clear ethical thinking about wild and domestic animals, as well as about holistic ecosystems. We are subject to the claims of close relationships, with people and with domestic animals such as pets, and also to "holistic environmental obligations" such as those articulated in Leopold's mandate to preserve the beauty, integrity, and stability of entire ecosystems, understood as the land.[19]

I am not sure whether Callicott is right that we need a coherent theory that can justify all the different ways we care about what we care about. I would like to bracket that issue as well and turn instead to a different question. Rather than thinking about our different ways of caring about animals and other aspects of nature alongside each other, as parallel philosophical streams, I will consider them in more direct relation to each other. I begin this task by noticing that the most persuasive attempts to find common ground between animal rights and environmental ethics, including those of Midgley, Callicott, and Fisher, focus on relationships. Their shared starting point is the recognition that we are in relationships, and communities, with animals, with other aspects of nature, and these relationships create moral sympathies, commitments, even duties. I ask what happens if we think about relationships even more deeply, and especially if we not only acknowledge but actively work on the relationships and sympathy that ground our moral commitments to the non-human world.

Sympathy for non-human nature is indeed natural, but it does not exist in a vacuum. There may be, and I think there is, an innate attraction between humans and other animals, plants, and places, whether we call it Wilsonian biophilia or Humean natural sympathy. It may, further, be strongest in children, as Midgley argues. However, it can be and in fact needs to be nurtured and cultivated—to be educated, like any natural desire, so that we desire "more and better."[20] If our natural sympathy for animals and nature is not encouraged and educated, it will not last. It may wither into indifference or be twisted into cruelty and destructiveness.

This gets us to perhaps the most important reason that environmental ethicists should take domestic animals seriously. In relationships with domestic animals, we come to recognize and then educate our natural desire for a mixed community

---

[17] *Ibid.*, p. 251.

[18] *Ibid.*, p. 256.

[19] *Ibid.*, p. 259.

[20] E. P. Thompson, *William Morris: Romantic to Revolutionary* (Stanford: Stanford University Press, 1981), pp. 790–1.

and cross-species bonds. In these interactions and relationships, we learn (morally and scientifically) appropriate ways to relate to non-human others. Ethics in this sense is a sort of "etiquette," as Jim Cheney and Anthony Weston have called it.[21] Ethical action is not an application of ideas we already figured out in the abstract but rather an effort to open up possibilities. In ethical action, we might learn to understand the dog or horse (and perhaps also the apple tree and tomato plant) as beings with their own ways of seeing and experiencing the world, with needs and interests that we can help fulfill or stymie, even as potential conversation and work partners. Working and playing with domestic animals is not just a process in which we enact already existing sympathies and values. Even more important, it is a process that itself generates and cultivates sympathies and values.

In this chapter I am thinking through this possibility in a very rough and preliminary way. My working hypothesis is that serious consideration of domestic animals can contribute to environmental ethics by focusing our attention on practices, relationships, community, and our own self-understanding as a species. Specifically, taking domestic animals seriously can help us recreate our relationships with nature, it can help recreate the meaning of community in and through our practices, and ultimately it can help recreate our self-understanding as ecologically and socially embedded creatures. Non-human animals, especially domestic species, serve as links between people and wildness and as facilitators of care for the non-human world. Pets and domestic animals, like children, help connect adult social worlds and wild worlds. The key is to respect the otherness of other animals while also finding doors between our different worlds that make positive relationships possible.

Recreate

God and the devil are both said to reside "in the details." Nature is also to be found in the details.[22] We do not know nature in the abstract, but in and through specific experiences and interactions. Environmental philosophers and activists often talk about the importance of direct experiences in nature for the development of positive environmental attitudes and behaviors. Wilderness activities such as backpacking, hiking, cross-country skiing, and kayaking take us into wide open spaces far from centers of human activity. Even the most dedicated outdoors-people among us, however, spend a relatively small fraction of our time in wild places. Overall, surveys show that today fewer Americans than in the past visit national parks, hike, backpack, cross-country ski, or otherwise get out in nature to have the kind

---

[21]   Jim Cheney and Anthony Weston, "Environmental Ethics as Environmental Etiquette," *Environmental Ethics* 21 (1999): pp. 115–34. See also Donna Haraway, *When Species Meet* (Minneapolis: University of Minnesota Press, 2008), p. 19.

[22]   See James Gustafson, *A Sense of the Divine: The Natural Environment from a Theocentric Perspective* (Cleveland: Pilgrim Press, 1994), especially pp. 13–14.

of sublime experiences that motivated Muir, Leopold, and other founding fathers of American environmentalism. If practical experiences are necessary for people to appreciate and try to protect nature, then we are probably in trouble. (We are in trouble anyway, but here is another possible reason for it.)

Fortunately, more mundane experiences with non-human nature can also be transcendent and life-changing. This is true because what is important is less the setting than the different way we act and interact when we spend time in places that are not structured primarily by human priorities. As people spend less time in wilderness, more and more of our experiences with nature are in back yards, empty lots, city parks, and other marginal places. These places, which blur the line between domestic and wild, provide opportunities for close interactions between humans and non-human nature.[23] It may even be that connections to nature and to non-human animals can be discovered and nurtured in partly domesticated places even better than in truly wild areas. Wendell Berry describes plowing a field while watched by a hawk who let him get close because he was with a team of horses rather than a noisy, smelly tractor. Amish farmer David Kline recounts a similar experience, one warm late summer day, when he snuck up on a fat woodchuck sleeping in the sun. "Taking my walking stick," Kline remembers, "I reached out and lightly scratched its back. Instead of waking, as I expected it to, the woodchuck arched its back in appreciation."[24] Kline's friendly encounter with a woodchuck, no less than a backcountry trek in Alaska, is an experience with and of non-human nature, of the sort that both reflects and changes our way of being in the world.

Like the edges of plowed fields, domestic animals are "boundary creatures," occupying a middle ground between wild nature and human society.[25] Still, for all their in-betweenness, domestic animals are part of non-human nature, never entirely beyond the influence of their wild origins. Encounters with domestic animals are the most frequent form of direct engagement with non-human nature for most people today, urban or rural. (Even more frequently, of course, people eat domestic animals, but that represents a form of indirect engagement.) Much of what we know and feel about nature comes from what we know and feel about domestic animals, to whom most people seem drawn from childhood on. Our sympathy for and attraction to nature may be innate, as the biophilia hypothesis suggests, but this sympathy needs to be cultivated in direct engagement and relationships. We can do that most fully with domestic animals, who are evolved to relate to us, and us to them.

Interactions with animals are not the same as ecological restoration or a hike in the woods, but they are still important and even necessary. Non-human animals,

---

[23]   Wendell Berry, *The Unsettling of America* (San Francisco: Sierra Club Books, 1977), p. 179.

[24]   David Kline, "Scratching the Woodchuck," in *Scratching the Woodchuck: Nature on an Amish Farm* (Athens, GA: University of Georgia Press, 1997), p. 145.

[25]   Michael Vincent McGinnis, "Boundary Creatures and Bounded Spaces," in *Bioregionalism*, ed. Michael V. McGinnis (London: Routledge, 1999), p. 61.

even highly domesticated ones like dogs, experience the world in distinctive ways, without the same goals, priorities, or assumptions of humans. When we try to meet these animals on their own terms, with appreciation for their agency and also their mystery, we experience non-human nature just as profoundly as when we gaze at charismatic landscape features such as towering mountains or crashing waterfalls—even more profoundly, perhaps, because our relationships with domestic animals can have a depth of communication that is not possible in many other encounters. One of the best ways to foster and deepen this communication is through recreation.[26]

Play with other species, as with other humans, takes many forms and enacts many meanings and values. I am interested in play that entails communication, greeting, and response; it has rules and standards, and ultimately it is about mutual respect and understanding. As I discuss below, in and through this type of play, we create relationships through play with another animal, whether in informal games of fetch or sustained, disciplined practices such as agility training with dogs or dressage with horses. This relationship is transformative, as is any real partnership.

When we play with other species, we create value, not unlike the process of ecological restoration that Bill Jordan and many other authors elsewhere in this volume describe. Sometimes we even create entirely new worlds. Certain kinds of relationships between humans and non-humans embody a different way of living and relating. In them we speak a second, often non-verbal, language, which can open a door to a different way of being. For Caroline Knapp, this door was opened by her relationship with her dog Lucille, who led her "into a world that is qualitatively different from the world of people, a place that can transform us. Fall in love with a dog, and in many ways you enter a new orbit, a universe that features not just new colors but new rituals, new rules, a new way of experiencing attachment."[27] Knapp recalls that she "once heard a woman who'd lost her dog say she felt as though a color were suddenly missing from her world: The dog had introduced to her field of vision some previously unavailable hue, and without the dog, that color was gone. That seemed to capture the experience of loving a dog with eminent simplicity. I'd amend it only slightly and say that if we are open to what they have to give us, dogs can introduce us to several colors, with names like *wildness* and *nurturance* and *trust* and *joy*."[28] Wildness, nurturance, trust, and joy can be experienced with people as well, but many aspects of human society can make them harder to find, or at least to recognize.

Joy, trust, and other positive aspects of mutual relationships between people and animals both make possible and emerge from play. In other words, we can play with animals because we acknowledge their agency but, equally, we can

---

[26]   We also do this through the domestication process, which itself "recreates" a species into something different than its wild counterpart. This is, of course a double-edged sword, with many far from benign consequences for non-human others.

[27]   Carolyn Knapp, *Pack of Two* (New York: Dial Press, 1998), p. 6.

[28]   *Ibid.*, p. 4.

acknowledge their agency because we play with them. In and through concrete practices, we come to understand the animals and ourselves differently. Mary Midgley observes that people who do well training animals do so by interacting socially with them, "by coming to understand things from their point of view." It is correct to speak of an animal's point of view, she adds, "because it means something to be a bat or a horse, in a way that is not true of machines."[29] If we cannot enter into another creature's point of view, if we continue only to "think like people," we cannot enter fully into the joy and partnership of interspecies play, and we will certainly never become "more than one but less than two."

Fortunately, playing with other species is not as hard as it sounds. As Midgley notes, "Play-signals penetrate species-barriers with perfect ease." Especially for young creatures of all species, including humans, the walls between species are full of holes.[30] In passing through some of these holes, we recreate ourselves and our understanding of other animals. Spending time with animals in loving relationships, notes Temple Grandin, teaches people "that there's more to animals than meets the eye."[31] There is more, in other words, than the passive, objectified, machine like images that dominate both scholarly and popular views of non-human animals. Interacting closely with animals in play makes it impossible to deny their agency, sentience, and moral status.

This brings us to the recreation of meaning made possible by recreation with non-human animals. Play can take us beyond compassion or pity, beyond Jeremy Bentham's famous question "Can they suffer?" to more promising questions, as Haraway suggests, such as "Can animals play? Or work? And even, can I learn to play with *this* cat. Can I, the philosopher, respond to an invitation or recognize one when it is offered? What if work and play, and not just pity, open up when the possibility of mutual response … is taken seriously as an everyday practice available to philosophy and to science?"[32] Everything changes when this possibility is taken seriously. Play does the most serious work in the world: it opens us to ways of being, to entire worlds, that we did not know existed—even ones that we strenuously denied could exist.

In the next section, I move from play to work, which turns out to be not such a clear line, and to the question of relationships more specifically.

---

[29] Mary Midgley, *Animals and Why They Matter* (Athens, GA: University of Georgia Press, 1983), p. 113.

[30] *Ibid.*, p. 117.

[31] Temple Grandin, with Catherine Johnson, *Animals in Translation: Using the Mysteries of Autism to Decode Animal Behavior* (New York: Simon and Schuster, 2005), p. 7.

[32] Haraway, *When Species Meet*, p. 22.

Relate

In play we build relationships, and depending on the kind of interactions we have with non-human animals, we can build more or less positive relationships, with more or less positive implications for our ways of being with non-human nature more generally. By learning how to relate to domestic animals in more mutual and respectful ways, we enact models for environmentally and socially responsible practices on a larger scale, recreating the meaning of community in specific practices and relationships. One way to do this is through a specific kind of play between humans and domestic animals, especially dogs and horses: the mutually transformative process of training for a particular discipline.

I am thinking of the kind of work, or play, that entails true partnerships across species. I do not mean to romanticize human relationships with domestic working animals, which have often been far from respectful or pleasant for the non-human partner. Still, in many cases, shared work is empowering, meaningful, and joyful for both humans and non-humans, who come to respect and rely on each other's distinctive capacities and agency. Dogs and horses are most often involved in this kind of work. It includes "professional" work such as that done by therapy, service, search and rescue, sniffer, and guide dogs and by working cow, farm, and therapy horses; it also includes play-work like agility training for dogs and equestrian disciplines such as dressage, among many others.

Donna Haraway has written extensively, in her most recent books, about agility work with dogs. This kind of work, she argues, is about communication, greeting, and response; it has rules and standards, and ultimately it is about mutual respect and understanding. While power differentials are always involved between different species, what is most important in the play Haraway describes is the dialogue of proposal and response, the ways that different bodies come together, "which makes each partner more than one but less than two." The transformative mutuality of play between different species might, Haraway contends, "underlie the possibility of morality and responsibility for and to one another in all of our undertakings at whatever webbed scales of time and space." This kind of recreation makes play joyful, not just "fun."[33] The joy comes in the reality of meeting another creature in mutual understanding, respect, and enjoyment. We create relationships through play with another animal, whether in informal games of fetch or sustained, disciplined practices such as agility training with dogs or dressage with horses. This relationship is transformative, as is any real partnership.

Play is not only a result of this mutuality but also, perhaps, its precondition. In other words, we can play with animals because we acknowledge their agency but, equally, we can acknowledge their agency because we play with them. In and through concrete practices, we come to understand the animals and ourselves differently. We may even stop thinking in species-bounded ways, as horse trainer Pat Parelli argues: "Most people are inadequate when it comes to horses because

---

[33]   *Ibid.*, pp. 242, 244.

they think like people. My goal is to get people to think like horses. The best way I know to do that is to play with horses on the ground,"[34] underlining the need to ground theory in practice. Knowledge about horses' evolutionary history is necessary but not adequate; information and theory cannot replace direct encounters and shared experiences that build and maintain any relationship. Haraway makes a similar point in writing about dog training. Good trainers, she contends, share a "focused attention to what the dogs are telling them, and so demanding of them … These thinkers attend to the dogs, in all these canines' situated complexity and particularity, as the unconditional demand of their relational practice." For the right sort of training, according to Haraway, "'method' is not what matters most among companion species; communication across irreducible difference is what matters."[35]

Both Haraway and Parelli emphasize the agency of non-human animals. As Parelli explains, "Communication is two or more individuals sharing and understanding an idea. If I pat my leg and the dog comes, we've communicated. But I can talk to a post until I'm blue in the face, and I'm just talking. Communication is a mutual affair between two or more individuals."[36] This exemplifies Midgley's argument, discussed above, that people who do well in training animals do so by interacting socially with them and coming to understand things from their point of view. Midgley reinforces a point central to the approach of Parelli and other trainers: to work successfully with non-human animals requires respect for the distinctive characteristics of each species and each individual, on the one hand, and willingness to find common ground, on the other.

In this disciplined play, we recreate the meaning of community in our practices and relationships. We move closer to what Midgley calls a "mixed community" that integrates human and non-human members. The mixed community begins with the fact of domestication, which occurs because animals form bonds with persons, as Midgley puts it, by coming to understand social signals addressed to them. This is possible "not only because the people taming them were social beings, but because [the animals] themselves were so as well."[37] The mixed community requires mutual communication, agency, and accountability. This kind of community differs greatly not only from the usual relations across species but also from the usual relationships between people.

---

[34] Pat Parelli, *Natural Horse*Man*Ship* (Colorado Springs: Western Horseman, 1993), p. 7.

[35] Donna Haraway, *The Companion Species Manifesto: Dogs, People, and Significant Otherness* (Chicago: Prickly Paradigm Press, 2003), p. 48 and p. 49.

[36] Parelli, *Natural Horse*Man*Ship*, p. 15. Parelli echoes a point made by Mary Midgley, who writes that the thrill of hunting comes from the relationship between the hunter and the prey animal who is their "opponent – a being like themselves in having its own emotions and interest." This is why, she concludes, shooting a rock is not a substitute (Midgley, *Animals and Why They Matter*, p. 16).

[37] Midgley, *Animals and Why They Matter*, p. 112.

This kind of relationship, and this sort of community, is built not only in and through play but also as a result of considerable hard work. We can learn a great deal about the relationship between work and community, I contend, by thinking about play with non-human animals, especially the sort of structured play that takes place in agility or dressage (or a host of other partnerships between humans and domestic animals). First, we learn that the line between work and play is a fine one. Through a lot of hard work, we might be lucky enough become "more than one but less than two" in relation to another creature. That is when work becomes play, or a dance between partners who fully understand and enjoy each other, and their mutual undertaking. Work becomes play also because it moves toward, and takes meaning from, common goals and values. This sort of work, and the play to which it is closely related, contributes to a larger community in multiple ways: it exemplifies proper relationships between people and among species, it values practices that are both useful and pleasurable, it seeks improvement through disciplined and creative practice, and it invites rather than excludes others.

This sort of work-play practice, and this sort of community, are of course far from common. It is very different from the commonplace ways we work, which divide us from other people, set us against non-human animals, and pit our private interests against the good of the whole. It does not entail participation in collective projects and goals, nor contributions to common goals, nor even realization of our own dreams. Work may seem to be an instrument to these goals, but work itself does not fulfill them. "Life itself appears only as a means to life."[38]

This kind of work is rarely confused with play, for one of the characteristics of play is that it is an end in itself, and not only a means to another objective. Work that is also play, such as the interspecies praxis I describe here, is an end in itself at the same time it is a means to other valuable goals—tangible accomplishments such as rounding up the sheep (or completing the agility course) as well as less tangible, equally desirable goals such as the strengthening of relationships, the cultivation of excellence, and the extension of understanding. Disciplined play-work between humans and companion animals can help "restore the wholeness of work," as Wendell Berry argues we must do. Berry proposes that "good work" should be not simply a means to an end, a way to maintain connections, but rather "the enactment of connections ... one of the forms and acts of love."[39] Good work, or what I am calling work-play, has its own intrinsic value in addition to, and regardless of, the instrumental role it plays in reaching practical ends. Just as in relationships among humans, the fact that practical goals are met need not make work alienating or relationships alienated. Even Kant did not require that we never treat others as means to our own ends—just that we do not treat them *merely* as means. Thinking about work and play together can help clarify the connections between ends and means, values and use, in more integrative and nuanced ways.

---

[38]   Karl Marx, "Economic and Philosophic Manuscripts of 1844," in *The Marx-Engels Reader*, ed. Robert Tucker (New York: W. W. Norton, 1978), p. 76.

[39]   Berry, *The Unsettling of America*, pp. 138–9.

Decenter

Throughout environmental philosophy runs a critique of anthropocentrism—the placement of humans as a species at the center of everything important—as a major obstacle to ecological sustainability. This critique seeks to replace human-centeredness with a view of humans as part of a larger whole. This claim is evident in a wide range of environmental philosophies, beginning with Aldo Leopold's land ethic, which changes the role of humans "from conqueror of the land-community to plain member and citizen of it."[40] A rejection of anthropocentrism is also foundational to Deep Ecology, which Arne Naess characterized as a more holistic view of humans' place in the world in contrast to the fragmentary and individualized view of "shallow" ecology, which instrumentalizes nature because it separates humans from it.[41] In an important early article, Naess argued that ecology suggests "a relational total field image [in which] organisms [are] knots in the biospherical net of intrinsic relations."[42] Other environmental philosophies similarly look to ecological science as a holistic model in which humans are but one among many mutually interacting organisms.

Many religiously-grounded environmental ethics also decenter humans, often by framing them as participants in an interdependent "web of creation." This more egalitarian or ecocentric ethic contrasts with the stewardship model, very common in monotheistic traditions, which separates humans from the rest of nature by assigning to them both special privileges and special responsibilities. Perhaps even more radically than the "web of creation" model is the theocentric approach of theologian James Gustafson, which shifts the center of value away from both humans and nature to God. Divinity, according to Gustafson, "has to include not only dependence upon nature for beauty and for sustenance, but also forces beyond human control which destroy each other and us. If God saw that the diversity God created was good, it was not *necessarily* good for humans and for all aspects of nature."[43] Gustafson's theocentric ethic decenters humans just as surely as do holistic ecocentric ethics—philosophical or religious. It may decenter them even more profoundly, further, than many ecocentric ethics, which often still allow room for human special privileges based on psychological, social, or evolutionary considerations. Gustafson's radical theocentrism offers transcendent grounds for relativizing and subordinating human interests and welfare, grounds which are not available to secular environmental ethics and which allow for little negotiation or

---

[40]   Aldo Leopold, "The Land Ethic," in *A Sand County Almanac* (New York: Ballantine Books, 1949), p. 240.

[41]   Arne Naess, *Ecology, Community and Lifestyle*, trans. and ed. David Rothenberg (Cambridge: Cambridge University Press, 1989).

[42]   Arne Naess, "The Shallow and the Deep, Long-Range Ecology Movements," *Inquiry* 16 (1973): p. 95.

[43]   Gustafson, *A Sense of the Divine*, p. 44.

compromise. Not all religious ethics offer such grounds, but Gustafson's model hints at the radical possibilities of a truly non-anthropocentric theological ethic.

Despite their important differences, these environmental ethics, both religious and secular, share a sense that humans must be decentered, their value relativized, if we are to appreciate the value of non-human nature. This decentering is very hard, perhaps impossible, to accomplish in theory alone. As environmental philosopher Anthony Weston puts it, "We should not suppose that we could *construct* a systematic non-anthropocentrism in the privacy of our studies or seminar rooms at all. Instead we must take up more systematically the entire question of the constitution of relationship in the first place."[44] We take up this question in and through a practical decentering: "We need to deanthropocentrize the world rather than, first and foremost, to develop and systematize non-anthropocentrism – for world and thought co-evolve. We can only create an appropriate non-anthropocentrism as we begin to build a progressively less anthropocentric world."[45] We must experience our own relativity, and the corresponding weight of other centers of value, in order to reduce our own importance in theory as well.

Weston proposes creating combustion-engine-free zones as an example of practical de-anthropocentrizing. Another, perhaps more immediately feasible experience, comes in and through concrete interactions with non-human nature—including domestic animals. Such experiences can help generate a sense of self that is more humble, more respectful of non-human others, more patient—all ecological virtues that are desperately needed in these trying times. Further, interactions with companion animals accomplish these goals in practice and not simply in theory. This is an important supplement or correction to much environmental philosophy, which focuses more on "getting the ideas right" than on praxis. Instead we need a more integrative understanding of ideas and practices, which continually constitute and transform each other.

Balancing the recognition of difference and otherness, equality and interdependence, is difficult and rarely achieved. It might become less daunting, suggests Mary Midgley, if we "get rid of the language of means and ends, and use instead that of part and whole. Man needs to form part of a whole much greater than himself, one in which other members excel him in innumerable ways. He is adapted to live in one. Without it, he feels imprisoned; the lid of the ego presses down on him."[46] This echoes a point made by E. O. Wilson and others: we cannot decide whether or not to be connected to other animals and ecosystems. Our relations to non-human nature are internal, part of our identities as individuals and as a species. Our choice is not about whether to be connected but about what to do with those connections, how to acknowledge and interpret them.

---

[44]   Anthony Weston, "Non-Anthropocentrism in a Thoroughly Anthropocentrized World," *The Trumpeter* 8, no. 3 (1991): p. 4.

[45]   *Ibid.*, p. 1.

[46]   Mary Midgley, *Beast and Man: The Roots of Human Nature* (New York: Routledge, 1995), p. 359.

Nonetheless, many people deny that they are part of a larger whole and that the other members of our whole, other creatures, are also agents, with their own ways of seeing and being in the world that are not, can never be, wholly determined by humans—no matter how hard we try, as in the case of many domestic animals. The objectification and instrumentalization of many non-human animals in the service of human food, labor, and entertainment is a moral, ecological, and political problem about which thousands of pages have been written. Deep moral and psychological, as well as ecological, contradictions reside in the huge divide we have perpetuated between the care for non-human animals and nature that most people profess, on the one hand, and the way these same people mostly treat the nature they encounter everyday, often without even recognizing it as such. In order to reduce this gap, and move toward valuing the intrinsic qualities of others—human and non-human—we have to engage in concrete practices with them. These practices are the foundation of better communities and better relationships, with people, with animals, and with non-human nature itself.

Bibliography

Berry, Wendell, *The Unsettling of America* (San Francisco: Sierra Club Books, 1977).

Callicott, J. Baird, "Animal Liberation: A Triangular Affair," *Environmental Ethics* 2 (1980): pp. 311–38.

Callicott, J. Baird, "Animal Liberation and Environmental Ethics: Back Together Again," in *The Animal Rights/Environmental Ethics Debate*, ed. Eugene C. Hargrove (Albany: SUNY Press, 1992), pp. 249–61.

Cheney, Jim and Anthony Weston, "Environmental Ethics as Environmental Etiquette," *Environmental Ethics* 21 (1999): pp. 115–34.

Fisher, John A., "Taking Sympathy Seriously: A Defense of Our Moral Psychology Toward Animals," in *The Animal Rights/Environmental Ethics Debate*, ed. Eugene C. Hargrove (Albany: SUNY Press, 1992), pp. 227–48.

Grandin, Temple, with Catherine Johnson, *Animals in Translation: Using the Mysteries of Autism to Decode Animal Behavior* (New York: Simon and Schuster, 2005).

Gustafson, James, *A Sense of the Divine: The Natural Environment from a Theocentric Perspective* (Cleveland: Pilgrim Press, 1994).

Haraway, Donna, *The Companion Species Manifesto: Dogs, People, and Significant Otherness* (Chicago: Prickly Paradigm Press, 2003).

Haraway, Donna, *When Species Meet* (Minneapolis: University of Minnesota Press, 2008).

Kheel, Marti, "The Liberation of Nature: A Circular Affair," *Environmental Ethics* 7 (Summer 1985): pp. 135–49.

Kline, David, "Scratching the Woodchuck," in *Scratching the Woodchuck: Nature on an Amish Farm* (Athens, GA: University of Georgia Press, 1997).

Knapp, Carolyn, *Pack of Two* (New York: Dial Press, 1998).

Leopold, Aldo, *A Sand County Almanac* (New York: Ballantine Books, 1949).

McGinnis, Michael V., "Boundary Creatures and Bounded Spaces," in *Bioregionalism*, ed. Michael V. McGinnis (London: Routledge, 1999), pp. 60–80.

Marx, Karl, *The Marx-Engels Reader*, ed. Robert Tucker (New York: W. W. Norton, 1978).

Midgley, Mary, *Animals and Why They Matter* (Athens: University of Georgia Press, 1983).

Midgley, Mary, *Beast and Man: The Roots of Human Nature* (New York: Routledge, 1995).

Naess, Arne, "The Shallow and the Deep, Long-Range Ecology Movements," *Inquiry* 16 (1973): pp. 95–100.

Naess, Arne, *Ecology, Community and Lifestyle*, trans. and ed. David Rothenberg (Cambridge: Cambridge University Press, 1989).

Norton, Bryan G., "Environmental Ethics and Nonhuman Rights," in *The Animal Rights/Environmental Ethics Debate: The Environmental Perspective*, ed. Eugene C. Hargrove (Albany: SUNY, 1992), pp. 71–94.

Parelli, Pat, *Natural Horse*Man*Ship* (Colorado Springs: Western Horseman, 1993).

Peterson, Anna, *Everyday Ethics and Social Change: The Education of Desire* (New York: Columbia University Press, 2009).

Taylor, Paul, *Respect for Nature: A Theory of Environmental Ethics* (Princeton: Princeton University Press, 1986).

Taylor, Paul, "The Ethics of Respect for Nature," in *The Animal Rights/Environmental Ethics Debate*, ed. Eugene C. Hargrove (Albany: SUNY Press, 1992), pp. 95–120.

Thompson, E. P., *William Morris: Romantic to Revolutionary* (Stanford: Stanford University Press, 1981).

Warren, Mary Anne, "The Rights of the Nonhuman World," in *The Animal Rights/Environmental Ethics Debate*, ed. Eugene C. Hargrove (Albany: SUNY Press, 1992), pp. 185–210.

Weston, Anthony, "Non-Anthropocentrism in a Thoroughly Anthropocentrized World," *The Trumpeter* 8 (1991): pp. 108–12.

Chapter 13

Replacing Animal Rights and Liberation Theories

Jonathan Parker

The word "replacing," as found in the title of this chapter, functions in a dual manner. One form of "replacing" entails substitution; that is, if I lose or break my phone, I replace it with a new one. As far as ethical theory goes, if we realize one theory is insufficient given the circumstances, we may seek a new theory to better address the situation. A second way to interpret "replacing" is in the sense of taking an object or concept and literally or figuratively putting it in another place. I argue in this chapter for the replacement of animal rights and liberation theories in this dual manner. That is, I do not wish to argue that these theories are in need of complete replacement in the former sense. The reason for this is that I do not make the claim that they are theoretically inadequate. Rather, I argue that these theories, which grew out of an initial concern for domesticated animals, are inappropriate and ultimately environmentally destructive if extended to cover our interactions with wild animals (and the interactions of wild animals amongst themselves). There has been a tendency for theorists to make this extension to encompass wild animals and nature, which I argue creates serious problems. If we restrict the scope of these theories however, and re-place them to only concern domesticated animals, they offer us sound frameworks for addressing our interactions with these animals in this context.

This chapter offers a connection and challenge to Anna Peterson's chapter in this volume (Chapter 12). Peterson suggests that domesticated animals are *liminal* beings, and that as our interactions with nature decreasingly constitute "wilderness experiences," our most frequent way of interacting with non-human nature is through domesticated animals. This is perhaps true for our conscious interactions (we do interact with wild nature all the time, though we may not perceive such interactions as wild since they often are not "wilderness experiences"), but even conceding Peterson's point, I contest that this fact is not very instructive in terms of how we ought to interact with wild non-human nature, as I do not think domesticated animals are as liminal as Peterson suggests. Domestication is a violent process which produces products of human culture; as such, human ethics (and as an extension, animal rights and liberation theories) adequately cover these cultured creatures. Suggesting domesticated animals are liminal creatures is potentially dangerous because it may suggest that our interactions with them teach us all we need to know about how we ought to interact with wild animals—which they do not. Extending our human ethics to cover domesticated animals is fair, but it is inappropriate for wild animals. Thus we are left with the need to replace these theories with more appropriate means of interacting with wild places and creatures.

Recreation

We are left with the need to put animal rights and liberation theories in their proper place, and consider appropriate wild ethics replacements. In other words, the differences between how humans interact with wild and domesticated animals in recreation provides a vantage point for examining the ethical differences between the two. We can begin with a discussion of "play," and the way Peterson deals with the topic highlights the problem. The types of play appropriate for encountering and coming to know domesticated creatures are extremely inappropriate for encountering and coming to know wild animals, and vice versa. We generally cannot encounter a wild animal and expect it to reciprocate in our play.

Alternatively, the types of recreation common for engaging wild animals—such as hunting—are fundamentally inappropriate for engaging domesticated animals. A longstanding method of recreating in nature, and engaging non-human animals was, and still is, through hunting and fishing. Animal rights and liberation advocates have historically fundamentally opposed this manner of interaction as cruel and inappropriate. Pioneering animal liberation and rights theorists Peter Singer and Tom Regan explicitly denounce hunting and call for its cessation in their groundbreaking texts. According to the ethical frameworks such theorists construct, hunting ought to be construed as an unethical method of interacting with wild animals and recreating in wild nature. Contrast this with José Ortega y Gasset, who argues that hunting is the ultimate means for humans to enter and recreate in nature, and the most appropriate and authentic means of encountering and interacting with wild animals.[1]

When we recreate with domesticated animals, killing them does not seem an appropriate means of interaction. Yet there is something very complex at the center of our interactions with wild animals. Here many would argue that engaging in the hunt is a central and ultimately appropriate and respectful means of interacting with wild animals. A great deal of tension arises when animal rights and liberation advocates confront the wild natural world with their theories. Peterson's treatment of domesticated animals as liminal beings may suggest something similar.

To get at some of the central issues at hand, it will be helpful to examine a recent paper from Ty Raterman which appeared in the journal *Environmental*

---

[1]   José Ortega y Gasset, *Meditations on Hunting*, trans. Howard B. Wescott (Belgrade, MT: Wilderness Adventures Press, 1995).

Ethics, entitled "An Environmentalist's Lament on Predation."[2] In this paper, Raterman, a self-described environmentalist, presents his personal lament on predation in nature, and argues that this does not undermine his status as an environmentalist. That is, he argues he can care for the environment, call himself an environmentalist, and still lament the fact of predation in nature. He devotes a short portion of his paper to a treatment of hunting, and I take it as implicit to his lament that human hunting as predation is ethically reprehensible. Sport hunting would be especially reprehensible, while subsistence hunting in his framework might be less so, and only lamentable.

Hunting as a practice draws attention to many of the central issues at stake in how we ought to interact with the natural world around us. Further, it is an interesting focus because our attitudes toward hunting are often naturally extended to our attitudes about the interactions of wild animals amongst themselves and in the wild generally. That is, in opposing hunting, animal rights and liberation proponents frequently slide from opposing hunting to opposing predation in general; such that we wish to stop the lynx from hunting the hare, etc. This move is a particularly problematic one when it occurs and ought to raise a red flag. I evaluate Raterman's position here, because he struggles with this very issue in his paper. It is necessary to evaluate the legitimacy and appropriateness of Raterman's lament and examine what effects such a lament has for interspecies ethics and whether this position is beneficial or dangerous. I focus on Raterman's paper, because it is illustrative of some of the troubles that animal rights and liberation theories run into when an attempt is made to extend them beyond our dealings with domesticated animals where, I would argue, such positions may be appropriate.

Lamenting Predation

Raterman states at the outset of his arguments, "That some animals need to prey on others in order to live is lamentable. While no one wants predators to die of starvation, a world in which no animal needed to prey on others would, in some meaningful sense, be a better world."[3] His reasons for finding predation lamentable are that it involves pain, it frustrates the prey animal's desires, and in general he thinks lamenting it is the most appropriate human response, and anything else such as celebrating or taking pleasure in predation is in conflict with virtues of compassion and gentleness. Paradoxically, and problematically, he appreciates and admires predator species for what they are and their various admirable attributes such as strength, speed, and agility, but regrets the process whereby they acquired those skills, i.e. through coevolution with the prey species in the hunt. He maintains that one can admire the product of a process without

---

[2]   Ty Raterman, "An Environmentalist's Lament on Predation," *Environmental Ethics* 30 (2008): pp. 417–34.

[3]   *Ibid.*, p. 417.

approving of the process itself. Finally, he fully admits that he is in some sense "opposing nature," but does not think that this disqualifies him from being an environmentalist, and claims that he can lament while not being committed to intervening in nature and disrupting predation.

What I think is important to emphasize is that Raterman begins his lament by testifying to having observed predation. He has witnessed predation live in the flesh; and not just any predation, but lions—the quintessential predator—hunting wildebeest in the Serengeti plains. It seems he thinks this fact does some work for him, or legitimizes his lament in some way. But let's note here how he witnessed this. He witnessed it, as far as the reader can tell, as a tourist on some kind of African safari, as a spectator safely distanced from the events. This fact is significant. I argue that he "sees" these events from a false spectator role that facilitates his admittedly anti-nature position, which leads him to oppose nature and the natural process of predation. Again, he admits predation is crucial to the development of the attributes of wild animals he loves so much, and along with human ecologist Paul Shepard, we could add that predation is crucial to the development of human intelligence which allows Raterman to write his lament. But again, Raterman wants to approve of the products of predation without affirming the process.

He gives the example of a recovered drug addict as a similar scenario where we might affirm the product—an individual strengthened by overcoming addiction—and yet lament the process. We can admire the strong character, and yet lament that it had to be forged through such hardship. Raterman only laments, maintaining that this lament does not logically commit him to intervening in and changing nature—which is a danger of extension that faces many animal rights and liberation advocates. So, while he would not seek the wholesale elimination of predation from nature, he does admit that he would find a world without predation to be a better world.

A final point worth addressing is Raterman's consideration of the effects his lament might have on human behavior. He argues we are not committed to stopping predation, but suggests that in cases where we can, perhaps we should. Significantly, he ends by saying humans should be called upon to be vegetarian by virtue of such a lament (which importantly he thinks should be the appropriate moral response of any decent human being).

This is where his lament has significant import. He claims that he is only lamenting, and as such, he is not committing himself to intervention or alteration of the natural order of things. However, he acknowledges that at least on a conceptual level, his lament leads him to oppose certain natural processes, and leads to significant lifestyle changes such as embracing vegetarianism. Raterman appears to smuggle in some weighty normative claims then when we acknowledge the consequences of such a lament, and combine them with the fact that Raterman thinks such a lament is the only appropriate human response to predation.

He claims that his intention at the outset was not to provide an argument for vegetarianism, but that this is, as it were, a byproduct of the lament. I suggest reaching such conclusions is a significant reflection of the internal logic of animal rights and liberation theories, and not just an interesting addendum. I pointed out at the beginning that there is a tendency to extend animal rights and liberation philosophies beyond the context of our relationship with domesticated animals, and here we see a problem that I think often occurs with animal rights and liberation theories, which is that if their logic is fully carried through, it can lead to these interventionist and anti-ecological positions that Raterman is trying so hard to avoid while still justifying the reasonableness of his lament.

Problematically, it appears that Raterman does not lament *all* predation. He tries to avoid the claim that many level against positions such as his, including Shepard, that such laments are simply anti-death. Raterman claims he is not anti-death, and gives some examples in which death could be a desirable thing and provide meaning to life more generally. He does not believe his vegetarian lifestyle counts as predation in any meaningful sense. He acknowledges that plants do strive for life, but finds no moral relevance to this striving. He says, "plants' 'striving' seems to be morally different from humans' and animals' desiring. I search my conscience and get no indication of what claim plants' striving makes on me (or on any human)."[4] He thinks that the claim animals make upon us is much more apparent. How could that be? One possibility is simply that certain animals are more like us, in that we perceive qualities in them that we ourselves possess and value. There is a long tradition in animal rights and liberation theory of extending rights outward based on similarity to the rights holders at the time. That is, we within the moral sphere may wish to grant moral consideration toward others based on qualities we ourselves value and claim to recognize, such as intelligence or the ability to feel pleasure or pain, etc. And the more like us other animals are, the more clear to us it appears that they have desires or can feel pleasure or pain, etc.

This is borne out by the peculiar feature that not all forms of interanimal predation seem to trouble Raterman. Remember, he testifies that he has witnessed lion predation of wildebeest. We can compare his response in that case with the following:

> I must concede, however, that not all animals clearly have desires. It is, for example, far from obvious that a fly has desires. The fly endeavors to stay alive, but doing so is perhaps on a par with the plant's striving; and so it is not clear that such a desire could be successfully appealed to in order to show that the spider's preying on the fly is lamentable. Indeed, it is partly on account of this, and also partly on account of the fact that there is no overwhelmingly good reason to believe that flies feel pain, that I do not particularly lament the spider's preying on the fly.[5]

---

4   *Ibid.*, p. 424.

5   *Ibid.*, p. 424.

This is unfortunately inconsistent; Raterman can witness with horror the act of predation among the mega fauna of the Serengeti plains, lamenting it and wishing it were otherwise, and yet shrug his shoulders, finding nothing morally relevant or lamentable at the *same process* occurring in his home when, for example, the spider ensnares his prey and devours it. The fly makes no demands of his gentle disposition, because as far as he can tell it's just trying to stay alive in the same sense as a plant, or at least it's not obvious to him that either has meaningful desires in the same way the wildebeest does. Though it is not entirely clear what makes him so confident about the wildebeest, it appears as though his lamentation is a gut reaction sparked by the plight of some animals but not others.

This type of lamentation is remarkably world denying. Raterman admits this to some extent; he realizes the apparent contradiction that he takes pleasure in many natural things and wild animals but laments that this world as it is must come to be through predation. Is there any value to such a lamentation? Perhaps this same world could have come to be without predation and in a future world predation could be eliminated from the framework, but that is not the world we live in.

If his lament did not have ethical and potential ecological consequences, perhaps we could allow him his grief, but since his lamentation commits us to alter our behavior and disposition to the natural world it deserves careful consideration and critique.

Of critical concern is whether these types of laments lead to intervention. Raterman argues his account does not entail that we should always, or even regularly, intervene in predation. Why? Quite surprisingly his "*first and most obvious reason*" is that it is too costly—in terms of consuming time, consuming energy, and consuming money.[6] This is not satisfying because it would seem that philosophically he is committed to ending predation, but acknowledges the futility of seeking to implement it in practice, though it would seem in his ideal world he would do so. Indeed he goes on to say that he does not want to suggest that we should "never, ever interfere in predation—only that [he] would be upset if [his] lament of predation entailed that we should always, or even regularly, do so."[7]

He gives further reasons, one of which he argues is more principled, where he maintains that there is no logical relationship between lamenting something and intervening. His example is a father who laments his son's engagement and choice of spouse, but nevertheless, out of respect for his son's autonomy, does not intervene. But this is not satisfactory either. Because again here he reveals his anti-nature sentiments but says out of respect for nature, he will not (always!) meddle, though ultimately he would wish to. My problem here is that it leads to a problematic spectator role. The logical consequences of animal rights and liberation theories, if extended to wild animals, are dangerous in one of two ways; either these theories imply a rejection of an ecological process that is central to the functioning of global ecosystems, which could lead to massive intervention to

---

⁶   *Ibid.*, p. 431.

⁷   *Ibid.*, p. 432.

reorganize ecosystemic processes and functioning, or they imply an ultimate and radical separation between human ethics and wild animal ethics, in the sense of letting those other creatures be. That is, we don't hunt, we don't predate, but we let animals act freely and we won't interfere. Suppose this were actually carried out, it would seem to end in a radically lamentable situation—our human behavior is improved by these standards, but the rest of the world flaunts us and our ethics, we still live in a lamentable world. How long could we hold out before ultimately extending our ethics to these lamentable predators?

The problem with positions such as Raterman's is they seem to push against ecological realities that life feeds upon life. Or as Aldo Leopold observed of a natural system, "The only certain truth is that its creatures must suck hard, live fast, and die often, lest its losses exceed its gains."[8] This may be rattling to those with gentle dispositions, but it is the world we live in. Raterman denies that he scorns this feature of life feeding on life and thus the centrality of death, but for Raterman the form of his denial of this claim seems beautifully utopian in the sense that he is accepting of death so long as it comes at the end of a fulfilled life for everyone and everything (except perhaps for plants and flies and others that he is not sure can lead meaningful lives, or have meaningful desires).

Conclusion

My point in this chapter has not been to reject animal rights and liberation philosophies. I would argue, along with people like Paul Shepard, that animal rights and liberation philosophies are important and good for those animals we have domesticated and brought into our homes and ways of life. In some sense it is right for them to fall under the rubric of our interpersonal ethics.

However we should stop short of extending this ethic to wild nature. Using the example of hunting as recreation, we have seen a need to stop short from using animal rights as a basis for our ethical relationship with wild nature. In fact, Raterman wants to stop short, but cannot seem to help himself from intervening. The way we recreate with domesticated animals is and ought to be different from the way we recreate with wild animals, and the one should not serve as a guidepost for the other, but all too often this is what happens in practice. There ought to be different forms of ethics circumscribing these very different relationships. For those animals that we have domesticated, we ought to come closer to having our interactions with them informed by an ethic similar to our interpersonal ethics, whereas this would be radically inappropriate for our interactions with wild animals.

What type of other ethic may we have available to us then? Shepard viewed the modern intellectual rejection of hunting as a rejection of the reality of "being a

---

[8]   Aldo Leopold, *A Sand County Almanac And Sketches Here and There* (New York: Oxford University Press, 1949), p. 107.

participant in a world where life lives on death."[9] It seems quite clear that Raterman laments living in such a world if not fully rejecting it. Following positions such as Raterman's, we often arrive at two options: we either embrace our lament while not intervening (most of the time), or the arguably worse alternative—we seek to restructure the world.

If we accept the more benign approach, then it seems we can only appreciate nature (if that is really appreciation at all) as an outsider; we venture out into it, but do not touch it. We observe the drama of the Serengeti from a safe distance, as we would a nature show on TV, but in this case a live performance. Shepard says of this, "When we try to extend our ethics to that with which it is incompatible, we get pictorial and esthetic images of nature, the Renaissance spectator, museum patronage, the culture of abstract appearances and dissociation."[10] He argues that such a view is a form of self-reproach; "It patronizes life, poses the important question of our true relationship to nature as condescension, and confuses a sentimental fiction with civilized enlightenment."[11]

It would seem we need something more world affirming, something that enters into the mystery of natural processes rather than denying them. And we cannot deny our participation anyway; observation from a distance is arguably a false outcome of a kind of nature/culture dualism. Shepard claims, "Bystanding is an illusion. Willy-nilly, everybody plays."[12]

If we follow Raterman then, an appropriate interspecies ethic would be non-interaction when it comes to human action, or an objectifying aesthetic observational relation, and a lamentable "letting be" when it comes to other animals interacting with each other and their surrounding environments. An alternative ethic would involve affirmation of wild nature and of ecosystemic processes. Hunting and predation in this context can be understood as an activity that, as Ortega says, "Submerges man deliberately in that formidable mystery and therefore contains something of religious rite and homage is paid to what is divine, transcendent, in the laws of Nature."[13] So one could say that Raterman, counting himself as an environmentalist, is a peculiar one indeed, characterizing himself as anti-natural, and excluding, in our interspecies ethics, any real interaction with wild animals. For people like Aldo Leopold, Paul Shepard, and Ortega y Gasset, hunting was a primary means of participating with and relating to wild animals. Ortega claims,

> Man cannot re-enter Nature except by temporarily rehabilitating that part of
> himself which is still an animal. And this in turn can be achieved only by placing

---

9   Paul Shepard, *The Others* (Washington, DC: Island Press, 1996), p. 317.
10   *Ibid.*, p. 317.
11   *Ibid.*, p. 317.
12   *Ibid.*, p. 319.
13   Ortega y Gasset, *Meditations*, p. 106.

himself in relation to another animal. But there is no animal, pure animal other than the wild one, and the relationship with him is the hunt.[14]

If this is the only real relationship we can have with wild animals, then Raterman does not need an interspecies ethic because he does not engage in relationships with wild animals.

Peterson asserts that the world which most of us participate in and encounter is increasingly a domesticated world. If this is true then an animal rights/liberation approach would be increasingly appropriate. This is partially true; however there is wildness all around us. Wildness should not be confused with wilderness. Wilderness is certainly diminishing, but not wildness. One simple definition of wildness is that which is not domesticated. Pet dogs are not wild animals then, because they have come to be as they are through the process of domestication, which alters their constitution and changes it from what it once was in the wild. Wolves on the contrary, not having successfully been through this process, are wild creatures. As with the spider in Raterman's house, there is wildness in our very homes. We need to be cognizant of these different forms of nature we encounter and participate in daily and realize that different contexts call for different modes of interaction and participation, and that ultimately these different contexts call for different ethical interactions.

What we must be most wary of is extending our animal rights and liberation views onto our interactions with wild animals and nature, for this can lead to world negating positions or calls for radical reconstruction of nature. Re-placed however, these theories are well suited to deal with our interactions with domesticated animals. Through our actions we have transformed domesticated animals into what they are and thus, I would argue, removed them from the wild community and brought them into our human communities. We have certain ethical duties to them as a result of this. It would be wrong, for instance, to hunt any domesticated animal that has, by virtue of being domesticated, lost its fear of humans and willingly approaches us. When authors like Tom Regan talk about the wrong involved in tricking or deceiving animals, this is certainly a good instance of reprehensible deception of an animal. This changes however when we look at wild nature where deception is a very common feature of many interactions among wild living things. What might an appropriate wild ethic look like? I think the workings of such an ethic are hinted at by ethical hunters such as Aldo Leopold and Ortega y Gasset, as well as the work of Paul Shepard: engaging and affirming wild animals and wild nature on its own terms and participating in it, as opposed to removing ourselves from it and denouncing its modes of interaction and natural processes.

An example of what actions this wild ethic would endorse may be beneficial at this point; to bring this chapter back to its starting point, let's take a recreation example. Like Raterman, I have my own safari experience, but quite a different one. On my honeymoon, my wife and I went out on an "aquasafari." We were given

---

[14]   *Ibid.*, p. 130.

helmets connected with an oxygen line, and the weight of the helmets pushed us to the sea floor where we could observe the local fish in their habitat. I was surprised when, as an incentive for the fish, our guide stuffed a net with white bread and affixed it to our helmets. Upon reaching the sea floor, the fish swarmed the bread. While this was certainly a good way of ensuring good viewing of the fish, and thus in some respects interacting with the wildlife, there was something unsettling about the experience. This seemed like an inappropriate way to be interacting with the fish. It is not that I thought we were harming the fish, but rather that this form of interaction seemed highly inauthentic.

Similar stories could be told about petting rays or swimming with dolphins. On one reading, this seems like a more animal friendly way of interacting with wildlife, and it is a move toward participating in the environment as distinct from being spectators. However, I think these actions are already steps toward domesticating these animals and cannot be praised as ethically appropriate ways of interacting with wildlife. There is a real sense in which a guided fishing expedition, following the guidelines of sportsmanship, is a more respectful and ethical way of interacting with the wildlife, particularly when the catch is consumed. In this respect, I am willing to grant that trophy hunting and fishing may be considered an inappropriate and unethical form of interaction with wild nature. In addition, embracing hunting and fishing as a means of ethically interacting with wildlife and providing sustenance, would perhaps ironically achieve the great reduction in animal suffering that people like Peter Singer desire. It is disturbing when people are willing to condemn hunting, yet consume factory-farmed meat. By confronting the food that we consume, we come closer to an understanding of where our food comes from, and how ecosystems function. Thus, while it is obviously unrealistic to expect everyone to hunt or fish for their own food, I argue that there is a great value in the experience of those activities and the types of interaction they require of the participant.

The sometimes-stark reality of the natural world and the way it functions poses a serious problem to our ethical sensitivities. I agree with Raterman that it ill befits a person of gentle disposition to embrace the realities of nature wholesale—given that there are many instances of predation that appear violent and cruel. However, we cannot withdraw from the natural world, and in some sense the horrors of factory farming were made possible by our cultural willingness to turn a blind eye to the realities of food production and supply. We should not be idealistic in either affirming or denying the natural processes of nature, but must seek a way to live well within the natural world as it is presented to us. I take this to be the challenge of developing a wild ethic. And while this wild ethic is something that will need to be treated further, I hope at least to have demonstrated the need to replace animal rights and liberation theories in the dual sense mentioned at the outset of the chapter. They are not helpful, and potentially are dangerous, if applied to the realm of wild nature and wild animals. However, if we can re-place them to only circumscribe our encounters and interactions with domesticated animals, these theories are nevertheless of great value.

Bibliography

Leopold, Aldo, *A Sand County Almanac And Sketches Here and There* (New York: Oxford University Press, 1949).

Ortega y Gasset, José, *Meditations on Hunting*, trans. Howard B. Wescott (Belgrade, MT: Wilderness Adventures Press, 1995).

Raterman, Ty, "An Environmentalist's Lament on Predation," *Environmental Ethics* 30 (2008): pp. 417–34.

Shepard, Paul, *The Others* (Washington, DC: Island Press, 1996).

Chapter 14

Re-Placing the Doctrine of the Trinity: Horizons, Violence, and Postmodern Christian Thought

Sarah Morice-Brubaker

A Christian community's understanding of divinity has profound implications for its understanding of environments, both built and natural. Consider, for example, the spatial and placial contrasts encoded in two theological models which are probably familiar to most readers: on one hand, the belief in a supremely simple, unconditioned, all-governing deity who is prior to both time and space but responsible for both; and on the other hand, the belief in multiple deities who act in time and reside within the same cosmos in which human beings reside.

To put the matter in such an abstract fashion brackets the influence of cultural factors—which arguably determine environmental ethics as much as do codified beliefs about divinity. And to be sure, this is hardly the only distinction which could be drawn between human models for imagining divinity. Nevertheless, I think it is an instructive one, because it shows so clearly how assumptions about environments can coincide with assumptions about divinity. Each of the schemas just mentioned, though they have to do primarily with theology, encodes possibilities about horizons and environments. Is divinity grounded inside of a horizon, or outside? Indeed, is it located at all? Or, rather, does being limited coincide with not being divine? What does it mean to be bounded? Is it the same as being contingent, as being less-ultimate? Is there an "outside" to the horizons (spatial, placial, historical, ontological) in which we find ourselves?[1]

In this chapter, I shall look at one particular tradition's understanding of divinity, as found in a particular religious tradition, at a particular philosophical and cultural moment, and expressed in the work of a particular contemporary theologian. Specifically, I shall be considering the Christian doctrine of the Trinity, in the contemporary Western theological milieu, as expressed in the work of the French philosopher Jean-Luc Marion. I say "philosopher," for this is how Marion

---

[1]   An excellent, exhaustive and invaluable treatment of this very broad topic is found in Edward S. Casey's influential work, *The Fate of Place: A Philosophical History* (Berkeley: University of California Press, 1998).

is most well known; but I shall be primarily concerned with his earliest, most explicitly theological work in which his thought is markedly Trinitarian.[2]

I choose this focus because I find it illustrative of a trend, both promising and troubling, in certain conversations within postmodern Christian theology. The classical doctrine of the Trinity has lately been used in attempts to solve philosophical problems that have cropped up in late modernity and postmodernity— namely, problems concerning God and horizons. Some Christian theologians, and particularly those who take Heidegger seriously, are lately scrambling to find a way to uncouple the quantity called "God" from some fixed and static ultimate category called "Being," without thereby rendering God entirely remote and unavailable. The doctrine of the Trinity is seen as offering promise here, due in part to its very inscrutability. Something about the structure of the Trinity (the thinking goes) yields a God who is not conditioned or fixed by Being, yet is not wholly absent.[3]

I said above that I find this trend both promising and troubling. The troubling aspect has to do with the Trinitarian model this strategy tends to yield. I shall parse this critique more carefully later, but briefly: although post-metaphysical theology is primarily concerned with refusing *ontological* horizons for God, at least in Marion's work the allergy seems to be triggered by *any* horizon—any

---

[2] Cyril O'Regan helpfully designates this portion of Marion's work as "theologically aspirated," by which he "mean[s] to mark that portion of Marion's work in which the discourses of the Christian tradition are read to mark an impossible opening beyond the regime of the self and the regime of metaphysics." O'Regan further offers this designation "without prejudice," even though Janicaud and others have rendered pejorative the outright description of Marion as "theological." Cyril O'Regan, "Jean-Luc Marion: Crossing Hegel," in Kevin Hart, ed., *Counter-Experiences: Reading Jean-Luc Marion* (South Bend: University of Notre Dame, 2007), p. 99.

[3] Catherine Mowry LaCugna, in *God For Us: The Trinity and Christian Life* (San Francisco: HarperSanFrancisco, 1991), uses the doctrine of the Trinity as a basis for a relational ontology, in the process explicitly criticizing the dominant ontology of the West which she understands to be over-focused on fixed, static, self-enclosed substances. William T. Placher, in his *The Triune God: An Essay in Postliberal Theology* (Louisville: Westminster John Knox, 2007), opens with a discussion of how "*start*[ing] with three" shields one against the Enlightenment/modern error of thinking of God as a discrete intra-mundane thing whose existence can be proven or disproven rationally. Karen Baker-Fletcher, in *Dancing With God: The Trinity from a Womanist Perspective* (St. Louis: Chalice, 2006), likewise uses the category of "dance" to describe the "communal, creative, loving, and relational" (p.11) activity of the immanent trinity. Again, part of Baker-Fletcher's aim here is to show God as something other than a static, discrete object of inquiry. Although there are other examples, I have chosen these three American theologians because they represent different perspectives within the theological academy—Rahnerian Catholic, postliberal Protestant, and womanist Protestant, respectively. Yet all three explicitly seek out Trinity as a corrective to theological approaches that would make God into an item within the metaphysical horizon.

backdrop against which God shows up, any economy in which God functions as a currency or concept, any set of conditions that God's revelation fulfills. We end up, rather, with a theology in which one of the advantages of the Trinity is that it gets God *outside of* and *prior to* all horizons and environments. Consequently, whatever disclaimers one builds in, it is still the case that "situated" gets mapped onto "not God" and "unsituated" gets mapped onto "God," and specifically onto "God *as triune.*"

I see both ethical and logical problems with this mapping. Logically, it is not clear how such a model allows for the advent of created place at all. As a category, place is utterly refused of the triune God—yet is already a facet of creation. Thus there is, so to speak, no "where" for environments or horizons to show up theologically. For it seems as though the only site where horizons could possibly be grounded is *between* the triune God (of whom any sort of horizon is refused) and creation (which is found already in a horizon). Yet the whole point of the theological model is to refuse any sort of "between," any sort of overarching category or horizon in which God and creation may each show up.[4] So while it makes sense, on Marion's own terms, that the recipients of God's self-gift would be thinking and speaking subjects, it makes little sense that they would be *placed* subjects. (And, perhaps not surprisingly, he says practically nothing about non-human creation.) But of course we are placed subjects, and there is quite a lot of creation that is non-human.

Ethically, I worry about the violent theological imagination that I see encoded in Marion's model. We end up with a God who is all-situating but not situated; displaced while bestowing place; delimiting but never captured; at once elsewhere and nowhere. Yet for all that, I actually do not think the idea of placing the three persons of the Trinity is an entirely bad one. And so in the last part of the chapter I will gesture toward some strategies to put place back *in* to Trinitarian thought, without reproducing the problems I see in Marion's model.

Trinitarian Theology and Place

In Christian theology—at least in its Western academic idiom—it no longer sounds strange to say that one's research interests focus on theology and place. Indeed, the last decade has seen a burgeoning interest in theologically-inflected inquiries into place, a function of the burgeoning interest in place across the humanities. To give but a few salient examples: Philip Sheldrake's *Spaces for the Sacred:*

---

[4]   I am here indebted to Lucy Gardner and David Moss' critical engagement with the place of the feminine in the work of Hans Urs von Balthasar, to whom Marion is indebted; in "Something Like Time; Something Like The Sexes – an Essay in Reception," in Lucy Gardner, David Moss, Ben Quash, and Graham Ward, eds, *Balthasar at the End of Modernity* (Edinburgh: T&T Clark, 1999), pp. 69–138.

Place, Memory, and Identity,[5] published in 2001, concerned the connection between constructed places and religious memory. A 2004 volume of essays from the Work Group on Constructive Theology is entitled *Spirit in the Cities: Searching for Soul in the Urban Landscape*. The book grew out of a series of meetings held in the late 1990s on "loci theology," in which participants reflected theologically on particular urban places: downtown Los Angeles; a US factory; a traffic corridor facilitating white flight in Newark, NJ; and an airport customs office, among others.[6] T. J. Gorringe considered the possibility of redemption for constructed spaces in his 2002 book *Theology of the Built Environment: Justice, Empowerment, Redemption*;[7] and this book was the topic for 2006 meeting of the Project on Lived Theology.

To say, though, that one has an interest in Trinity and place, is still to say something a bit idiosyncratic. Theological reflection on the Christian doctrine of the Trinity—the belief that God is both three and one—has often proceeded with the proviso that, of course, we're not talking about *corporeal* realities; and hence, not as though there are three discrete quantities somewhere, bounded and located in the same manner as three objects.

This qualification is particularly understandable given the matrix in which the doctrine first emerged. Trinitarian thought first developed within a relatively new religion that had to give a nod to the monotheism of its Jewish roots, as well as to the philosophical commitments of its Hellenistic roots.[8] Neither the Jewish proscription of idolatry, nor the Platonist pessimism about corporeality, could have much patience for notions of God being either plural or placed. Thus, developing Christianity had to walk a fine line. The careful negotiation began when influential streams of Christianity affirmed that their central figure, Jesus, was at once the messiah promised by the Jewish deity, and the eternal *Logos* of Hellenistic philosophy (albeit startlingly enfleshed). It continued when Christian communities had to wrestle with the "Spirit" language in their scripture and liturgy—language which seemed to ascribe some kind of divinity to the Spirit, even while distinguishing the Spirit from Jesus and from the one whom Jesus called Father. And it became, finally, formalized in the doctrine of the Trinity. In its dominant expressions, the doctrine affirms *both* that God is Father, Son, and Spirit,

---

[5]   Philip Sheldrake, *Spaces for the Sacred: Place, Memory and Identity* (Baltimore: The Johns Hopkins Press, 2001).

[6]   Kathryn Tanner, ed., *Spirit in the Cities: Searching for Soul in the Urban Landscape* (Minneapolis: Fortress Press, 2004) pp. xii–xiii.

[7]   T. J. Gorringe, *A Theology of the Built Environment: Justice, Empowerment, Redemption* (Cambridge: Cambridge University Press, 2002).

[8]   For a helpful summary of this doctrinal development, see Jaroslav Pelikan, *The Christian Tradition: A History of the Development of Doctrine*, vol. 1, *The Emergence of the Catholic Tradition (100–600)* (Chicago and London: University of Chicago Press, 1971), pp. 172ff.

with each a distinct divine Person rather than simply modes of divine expression; *and* that God is one, single, undivided and unconditioned.

Granted, in classical Trinitarian thought, there are some gestures that could be seen as placially significant, with terms like *perichoresis* ("interpenetration") and *oikonomia* ("household" or "economy") being adopted as metaphors for the relationship between the three divine persons. There are also some theological outliers, like the second century Tertullian, who seem to ascribe some kind of corporeality to the three persons—though in Tertullian's case he means to juxtapose "corporeal" with "null" or "nonexistent."[9] Yet the dominant strand of classical Christian Trinitarian thought proceeds with constant reminders that we are not dealing with three bounded, spatially located quantities. Basil of Caesarea, in the fourth century, marvels at "the stupidity of confining incorporeal beings in defined places."[10] Arguably, his opponents were doing no such thing; but the smear itself was rhetorically effective. Gregory of Nyssa, Augustine, Thomas Aquinas, and other architects of the classical (in a broad sense) notion of the Trinity rehearse similar disclaimers.

A New Development in Trinitarian Theology?

Perhaps this is why the contemporary Christian theological academy, having inherited this tradition, is not in the habit of considering theology of place *alongside* Trinitarian theology, even though both are active areas of reflection. So when I suggest that place language may be creeping into contemporary Trinitarian thought, I do not mean to suggest that this has happened in an especially programmatic way. By and large, theology of place is one conversation, and Trinitarian theology is another.

Still, over the last several decades, I believe that one can detect the beginnings of a mutual influence between the two conversations. And in the case of Jean-Luc Marion's early, more theologically-inflected work in particular, place categories— "distance," "site"—take on great structural importance for the Trinitarian model he suggests. Surprisingly, though, these placially-inflected words function exactly as a placial *refusal*—as a way of saying that God, qua triune, is in no sense placed; while everything that isn't God is placed.

As I mentioned above, this maneuver must be understood as part of a contemporary debate about God and ontological horizons. The contours of this problem, as laid out so devastatingly by Heidegger, have to do with an alleged confusion of God and Being that has plagued Western thought.[11] A fixed quantity

---

[9]   Tertullian, *Against Praxeas*, VII.

[10]   Saint Basil (Bishop of Caesarea) *On the Holy Spirit* 6.15, David Anderson (trans.), Popular Patristics Series (Crestwood: Saint Vladimir Seminary Press), p. 30.

[11]   This rehearsal is a main focus of Marion's *God Without Being*, Thomas A. Carlson (trans.) (Chicago: University of Chicago Press, 1991). Also, on the influence of Heidegger's

called "God" has, according to this view, been offered as both the source of Being and Being itself. Aside from the fact that this maneuver bespeaks a certain confusion—i.e. is God conditioning and situating Being, or vice versa?—it is also said to reflect a totalizing, violent impulse that seeks to erase difference by fixing it conceptually within the master horizon of Being.

Marion is a reader of Heidegger—as well as of Levinas, Derrida, Nietzsche, Altizer, and other continental philosophers who (albeit in different ways) argue that to fix something in a concept is to exert mastery over it, to dominate it, to collapse its otherness into a totalizing sameness ... in sum, to make it an item within our own horizon. If true, this would clearly be fatal for theology.[12] Theology consists, of course, of words about God; but if conceptualizing God means stripping God of God's alterity, making God into just another unit in our shared economy of signs and words, then what sort of a God are we left with? Nothing, it seems, but a God of our own making, a dead God, an idol. But the alternative—excluding the possibility of any words or concepts about God—bodes no better for theological discourse.

Can there be words about God? Can God show up within our horizons and economies, in ways that still preserve God's alterity? Marion's answer is, essentially, yes—*if* the God we're talking about includes, in that God's very self-definition, a refusal of horizons. If such a refusal is built in to the doctrine of God, then God can be revealed *within* a horizon without being fixed in that horizon. This refusal, to use one of Marion's own images in his Levinas-inspired essay "The Intentionality of Love," is rather like the pupil in the eye of the one whom you behold. In beholding an Other, I am tempted to reduce the Other to simply a fixture or object located in my gaze, an assemblage of traits and qualities whose value lies in their comprehensibility and availability to me. The Other's pupil, though, designates that which refuses this kind of reduction. In "this ever black point," Marion writes, "in the very midst of the visible, there is nothing to see, except an invisible and untargetable void."[13] And this tiny void, of course, is also the occasion for my feeling the weight of the Other's gaze *on me*: "my gaze, for the first time, sees an invisible gaze that sees it."[14] Neither the Other, nor my being gazed upon by the Other, are within my control. Nor can I be entirely accounted

---

criticisms of Western metaphysics upon Marion, see Robyn Horner, *Jean-Luc Marion: A Theo-logical Introduction* (Aldershot: Ashgate, 2005), esp. pp. 35–46.

[12]   For perhaps the most illustrative example of Marion's estimation of the totalizing nature of metaphysics, particularly as regards place, see his tantalizingly placially-titled, "The Marches of Metaphysics," the opening essay in Marion, *The Idol and Distance: Five Studies*, Thomas A. Carlson (trans.), Perspectives in Continental Philosophy, John D. Caputo, ed. (New York: Fordham University Press, 2001).

[13]   Jean-Luc Marion, "The Intentionality of Love," in Marion, *Prolegomena to Charity*, Stephen Lewis (trans.) (New York: Fordham University Press, 2002), p. 81; originally published as *Prolégomènes à la charité* (Paris: E.L.A. La Différence, 1986).

[14]   Marion, "The Intentionality of Love," p. 81.

for by the Other, for I too have a pupil, and a gaze. In this crossing of gazes, the irreducibility runs in two directions. This irreducibility protects each party from ultimate objectification in a fixed concept, but it also smashes each party's pretense to objectivity: I discover myself, in a kind of reversal, as a gazer who is gazed upon.

Earlier I suggested that if Marion is to recuperate theological discourse, he will need to explain how God is neither entirely unknowable to, nor entirely reducible by, human conceptual knowledge. In "The Intentionality of Love," Marion has offered the pupil as a marker of irreducible alterity in the crossing of gazes. This, though, is a phenomenological model, not an explicitly theological one. But when Marion makes the theological move, he likewise takes as one of his primary tasks the designation of something in God that refuses all reduction and recuperation. That something is *distance*—specifically, the eternal distance between the Father and the Son in the Trinity.

We ought not overlook how startling and paradoxical a move this is. "Paternal distance," writes Marion, "offers the sole space for a filiation."[15] "Distance" is, ordinarily, a placial term. It usually implies a discrete interval or extension and, it follows, some prior matrix or backdrop or horizon or set of coordinates *within which* that interval can be plotted. Here, though, "distance" names precisely the *refusal* of any prior horizon or set of coordinates, including Being. Moreover, that refusal is built into the very structure of the Christian doctrine of the Trinity. The idea, if I understand Marion correctly, is that the Father ever withdraws into an unbridgeable, unspannable, irrecuperable distance with no terminus. Or perhaps it is best imagined by way of contrast: if totalizing knowledge involves a move closer—a drawing near to something in order to gaze upon it, reduce it, objectify it, and render it a static fixed quantity to be traded in the semiotic economy—then distance entails precisely the opposite movement. Distance involves a flight away from, an evacuation, a withdrawal, a retreat into absence, a turning-away, a refusal to be manifest.

If this sounds as though it comes precariously close to atheism, that is partly Marion's point. Marion takes seriously Nietzsche's proclamation of the "death of God," though Marion holds that it is a particular sort of "god" who has died.[16] Reading Marion, we are meant to keep in mind a more familiar template for divinity—one in which a deity straightforwardly shows up, dazzling us with

---

[15]   Marion, *The Idol and Distance*, p. 139.

[16]   See, for example, Marion's discussion of Nietzsche in Chapter 2 of *God Without Being*, in which Marion portrays the will to power as consonant with, rather than a departure from, the idolatry of metaphysics. "Just as we were able to venture that Nietzsche, because he carries metaphysics to completion, constitutes its last moment," Marion asserts, "so must we suggest that Nietzsche renders the twilight of the idols crucial only by himself consummating a new (final?) development of the idolatrous process. The will to power forges 'gods' at every instant: there is nothing, in the modern sense, more banal than a 'god'" (p. 38).

splendor according to the terms of our own horizons, offering a manifestation which precisely fulfills our conditions for divinity. Indeed, for Marion, the chief distinction between God and an idol is that an idol obligingly shows up as present for all to see, whereas God is—in a very real sense—absent to sight and thought. For Marion, those who cry "God is dead!" come very close to the truth. For the atheist looks and, quite correctly, sees that God is absent.[17]

What atheism fails to perceive, on Marion's view, is that there is a divine presence coinciding with the very real divine absence—and that this absence regards *us*. Here again, the argument is a Trinitarian one. Where "Father," for Marion, designates God's withdrawal into irrecuperable distance, "Son" designates the divine disclosure contained therein. Formally, this recalls the structure of Marion's argument in "The Intentionality of Love." The pupil, we recall, marks the Other as irreducible to the terms of my gaze, while also reversing the direction of the gaze so that I experience myself as gazed-upon. In a formally similar manner, the Son marks the Father as withdrawing, as distant, as exceeding our horizons and conditions ... above all, *as an Other* who gazes. Here, too, is the reversal: "Distance is not given to be understood," Marion quips, "since it understands us."[18]

Indeed, precisely *because* the Father withdraws into distance—prior to and outside of time or space or even Being—the Father thereby makes alterity possible. There can be an Other to the Father, who feels the weight of the Father's gaze, exactly *because* the Father disappears into distance. This theological rule, it is crucial to understand, operates on two levels. *Within* the Trinity, such alterity is exercised by the Son. The Son and the Father gaze at each other in love, across a yawning abyss of distance, neither placing any reductive claim upon the other. (This, on Marion's reading, is the mystery revealed in time and space, in

---

[17]   Discussion of Nietzsche's influence upon Marion—and particularly his indebtedness to Nietzsche's notion of God's death—raises the related question: what of Hegel, who also, albeit differently than Nietzsche, is often understood to have collapsed God into intra-mundane immanence? O'Regan discusses the surprising feature of Hegel's role in Marion's theologically-aspirated texts: "[O]ne notices something, or at least one thinks one does. Or rather one notices a nothing, an absence rather than a something. One does not come across Hegel; one does not see Marion crossing in front of, through, or over Hegel, as he so obviously does with Husserl and Heidegger. Why the nonappearance, the failure to show[?]" (O'Regan, "Jean-Luc Marion: Crossing Hegel," p. 95). O'Regan concludes that the conspicuous disappearance of Hegel in fact constitutes Marion's fierce resistance to, and attempted marginalization of, Hegel. In Marion's theologically aspirated works, Hegel's pneumatology is implied to be especially unacceptable. For Marion, "Christ ... establishes the parenthesis within which existence and history happen. But if instead of distance there is dialectically overcomeable alienation, then filiation is put in question ... Christ becomes not simply contingently replaceable, but necessarily so, if one gives in to the seduction of a Hegelian-style pneumatology." O'Regan, "Jean-Luc Marion: Crossing Hegel," pp. 102–3.

[18]   Marion, *Idol and Distance*, p. 153.

the Father's silence as Jesus was dying on the cross.) But there is a "crossing," as well, *between* the triune God and creation. Indeed, for Marion, the distance between Father and Son actually grounds—in a derivative and secondary way— the relationship between God and creation; or as Marion calls it, the "crossing of Being."

It is a brilliant theological maneuver, inasmuch as it deftly addresses the objections from Marion's philosophical interlocutors. According to Marion's model, the inner life of the Trinity *already* includes absence and presence, alterity and unity, and love across unbridgeable distance. There is not some prior theological or ontological horizon which demands that these processes be included in the Trinity. Rather, these processes are at play within the Trinity exactly *because* the triune God is prior to all horizons. It is this dynamic interplay—of presence coinciding with absence, alterity coinciding with unity—which gives us "Father" and "Son." Paternal-filial distance is not (it turns out) a set of operators drawn solely from within our horizon, a set of conditions which God obligingly fulfills. Rather, paternal-filial distance is at once the *conditions of possibility* of the divine disclosure; the *content* of the divine disclosure; and the very *divine persons* who are revealed, but never conceptually captured. To behold this disclosure is to experience oneself as gazed-upon, as created, as one whose very being is sheer gift.

Problems and Possibilities

I write as a theologian who is interested in connecting current theological conversations about place with critically informed appraisals of classical doctrines of God. As such, I see both possibilities and problems in Marion's project.

To begin with the problems: Marion's rejoinder to his philosophical interlocutors is adroit and ingenious, but nevertheless I detect two logical tensions in his overall argument. The first, as I argued in the introduction, has to do with the status of horizons. Why are there horizons? Put differently, why are horizons—environments, places, spaces, matrixes, habitats, worlds, locations, etc.—a hallmark of created life? When did they enter the picture? For in Marion's model, *prior* to time and space and Being, is none but the triune God who refuses all horizons. Yet on the created side of the divide, everything is *already* located. Clearly horizons did not originate in God, but they occupy a priority within creation that Marion leaves unexplained. In this respect, placed-ness is different from other aspects of human life—existence, subjectivity, conceptual thought— which Marion does take care to ground in Trinitarian logic. So why are horizons deprived of theological status?

A second problem emerges when we consider one element of Marion's Trinitarian thought which is, to put it mildly, diminished. Where is the Holy Spirit? I would grossly overstate the case if I were to suggest that Marion never mentions the Spirit. However, I think it is fair to say that the Spirit does not occupy nearly the structural importance in Marion's model as do the Father and the Son.[19] Indeed, it is difficult to see how the Spirit *could* occupy a status equal to the Father and the Son, given how thoroughly Marion allows the same/Other binary to situate his thought. "Son" exactly names alterity from the Father, and in so doing, exhausts all possibility for distance from the Father. All other distance— such as the distance between God and creation, as we have seen—derives from this anterior distance between Father and Son. Framed this way, how could the Spirit have its own alterity, of equal status to the Son's? How could we help but have a binarity instead?

In addition to these internal problems, I am concerned about some of the ethical implications of Marion's project due to the vast reach he gives to theological discourse, and horizons' lack of any theological status. In classical Trinitarian reflection,[20] the second person of the Trinity is not only referred to as Son, begotten of the Father … but also (in a nod to Christianity's Hellenistic origins) as the Word/*Logos*, spoken by the Father. In Marion's work there is a parallel between this divine Word, and theological words about God. Just as, for Marion, the distance in the heart of the Trinity is all-situating but not situated; likewise, theological discourse is given unconfined scope. Or nearly so: Marion's one stipulation is that theological discourse needs to take care not to be idolatrous, and so cannot claim to fix God within a concept. Other than this, theology need not answer to any other discourse.

Certainly, one could take a harsher or softer view of the reach which Marion gives to theological discourse. One could argue that since he makes theology non-predicative—so that theology properly renders praise *to* God rather than making self-styled objective claims *about* God—that Marion's model possesses an appropriate kind of modesty, a built-in hedge against epistemic "violence." But when this element of Marion's thought is considered alongside other elements, it would seem, a different picture emerges. Even with Marion's provisos, one of his main polemical aims is to have theology situate metaphysics rather than the other way around, in a manner similar to, and grounded in, the way that the distance between Father and Son situates creation. If the analogy works, it works because theology is unsituated with respect to other discourses. Does this not substitute one violent and totalizing discourse for another? Marion, I expect, would reply that totalizing discourses try to fix, to nail down, to objectify, to make static; whereas

---

[19]   See n. 17 (above) and the entirety of O'Regan, "Jean-Luc Marion: Crossing Hegel." The tendency of Marion to erase the Spirit from the Trinity may be a function of his concerns regarding Hegelian pneumatology.

[20]   *Logos* Christology predates dogmatic articulations of the Trinity; it is evident in the Gospel of John, for example. I have in mind here the second through fifth century trajectory of Trinitarian reflection which includes such figures as Justin Martyr; Irenaeus of Lyon; Clement of Alexandria; Origen; Athanasius; the Cappadocian fathers (Gregory of Nyssa, Gregory of Nazianzus, and Basil of Caesarea); and Augustine of Hippo. See William G. Rusch, ed., *The Trinitarian Controversy*, Sources of Early Christian Thought (Philadelphia: Fortress, 1980).

theology endlessly defers and refuses to objectify. I am not so sure. Marion's version of Christian theology may include endless deferrals and refusals to nail down, but it also, and perhaps for that very reason, denies any horizons for anything other than itself. Whatever difference there ends up being, it is thoroughly situated by Christian theology: exactly because there is no possibility for difference other than Christian Trinitarian difference. It may be that nothing gets nailed down, but it is also true that everything is theological. Is this really less totalizing? Instead of the conquering horizon of Being which collapses all difference into itself, we seem instead to have an all-conquering Word which, by marking irreducible distance, situates absolutely everything. Given how the dissemination of Christianity has so often advanced through violence and colonialism, ought not Christian theology be more cautious here? (To my knowledge Marion has thus far been impervious to this criticism.)

Let us hold on to that general criticism, while also raising a subsidiary point about Marion's allergy to horizons. When theological discourse is here afforded such an elevated position, *and* environments have no theological status, then environmental violence is inaccessible to Christian theological critique. Indeed, it would seem as though *any* environmental issues would become inaccessible to Christian theology. This may not be Marion's concern, but it is nonetheless worth noting that he still takes certain possibilities off the table: environmental theology, intersections of theology and geography, and theological accounts of sacred spaces, to name but a few examples.

In sum, even though the placially-resonant term "distance" is Marion's central Trinitarian category, we do not actually end up with a Trinitarian theology of place. Yet for all that, I do see some very promising aspects of Marion's project. For a start, there is his choice of interlocutors. Any theological inquiry into place will have to contend with the philosophical problem of how the absolute can be present within a horizon. Marion takes this as his central polemical aim, to his project's good.

Moreover, by attending to the strengths and weaknesses that attach to Marion's understanding of distance, we can begin to imagine the features of an alternative term. One shortcoming, as we have seen, is the way in which distance—even irrecuperable distance—tends to suggest two poles: a same and an Other, the Father and the Son. Again, if it is important to preserve Trinitarian conventions, then the Spirit must be added to the picture. And for the sake of logical completion and consistency, horizons must be given some sort of theological status. Why not make both amendments in one theological move? For "distance," substitute "place." The objection will of course follow that one cannot have God be exhaustively conditioned by extra-divine horizons. But surely an answer suggests itself. In Marion's model, "distance" maps onto the Father and the Son. Following Marion's lead, if not his example, we could plausibly say that the Spirit is the "place" which the paternal-filial distance spans.[21]

Were one to make this amendment to Marion's project, I believe, many other theological pieces would fall into place. It would be possible, for example, to imagine created horizons as being grounded in the "where" of the Trinity, the Holy Spirit. This would create an opportunity to construct a Christian environmental theology which proceeds in a way that attends to its implicit claims about divinity. And of course the influence would run in the opposite direction as well: the Trinitarian model that results would have to take seriously its implications for environments, places, and horizons. Moreover, it would have to proceed with an appreciation of its own contextuality, at least if, following Marion's model, the rules governing theological discourse derive from the structure of the Trinity. Here, the divine Word is always a Word *in place*; and so, too, will theological words be words *in place*.

To be sure, as a theological model, what I am suggesting only has a chance, initially at least, of satisfying those with very particular interests—namely, post-metaphysical theology conducted in a Christian, Trinitarian idiom. But I am not offering it as an authoritative model—let alone as the final word on the doctrine of the Trinity—so much as a starting point for further work and conversation. Marion's model, as I understand it, forbids deep engagement with environmental philosophy and theology, postcolonial thought, theology of religions, and contextual theology. The amendments I propose may cause more problems than they solve. They do, however, stand to bring more parties to the conversation, including interlocutors who are well-equipped to expose some of the more violent aspects of the Christian theological imagination. In my opinion, this can only be to the benefit of Christian theology, as it seeks to maintain its coherence as a discourse without proceeding in a violent, totalizing fashion.

Bibliography

Baker-Fletcher, Karen, *Dancing With God: The Trinity From a Womanist Perspective* (St. Louis: Chalice, 2006).

Basil (Bishop of Caesarea), Saint, *On the Holy Spirit*, David Anderson (trans.), Popular Patristics Series (Crestwood: Saint Vladimir Seminary Press, 1997).

Casey, Edward, *The Fate of Place: A Philosophical History* (Berkeley: University of California Press, 1998).

Gardner, Lucy and David Moss, "Something Like Time, Something Like the Sexes – an Essay in Reception," in Lucy Gardner, David Moss, Ben Quash, and Graham Ward (eds), *Balthasar at the End of Modernity* (Edinburgh: T&T Clark, 1999).

---

[21]   This is the model I advance in my dissertation, *The Place of the Spirit: A Trinitarian Theology of Location* (University of Notre Dame).

Gorringe, T. J., *A Theology of the Built Environment: Justice, Empowerment, Redemption* (Cambridge: Cambridge University Press, 2002).

Horner, Robyn, *Jean-Luc Marion: A Theo-logical Introduction* (Aldershot: Ashgate, 2005).

LaCugna, Catherine Mowry, *God For Us: The Trinity and Christian Life* (San Francisco: HarperSanFrancisco, 1991).

Marion, Jean-Luc, *God Without Being*, Thomas A. Carlson (trans.) (Chicago: University of Chicago Press, 1991). Originally published as *Dieu sans l'être: Hors-texte* (Paris: Librairie Arthème Fayard, 1982).

Marion, Jean-Luc, *The Idol and Distance: Five Studies*, John D. Caputo (ed.) (New York: Fordham University Press, 2001). Originally published as *L'idole et la distance* (Paris: Editions Bernard Gasset, 1977).

Marion, Jean-Luc, *Prolegomena to Charity*, Stephen Lewis (trans.) (New York: Fordham University Press, 2008). Originally published as *Prolégomènes à la charité* (Paris: E.L.A. La Différence, 1986).

O'Regan, Cyril, "Jean-Luc Marion: Crossing Hegel," in Kevin Hart (ed.), *Counter-Experiences: Reading Jean-Luc Marion* (South Bend: University of Notre Dame, 2007).

Pelikan, Jaroslav, *The Christian Tradition: A History of the Development of Doctrine*, vol. 1, *The Emergence of the Catholic Tradition (100–600)* (Chicago and London: University of Chicago Press, 1971).

Placher, William, *The Triune God: An Essay in Postliberal Theology* (Louisville: Westminster John Knox, 2007).

Rusch, William (ed.), *The Trinitarian Controversy*, Sources of Early Christian Thought (Philadelphia: Fortress, 1980).

Sheldrake, Philip, *Spaces for the Sacred: Place, Memory and Identity* (Baltimore: The Johns Hopkins Press, 2001).

Tanner, Kathryn (ed.), *Spirit in the Cities: Searching for Soul in an Urban Landscape* (Minneapolis: Fortress Press, 2004).

Tertullian, *Against Praxeas*, Philip Schaff (trans.) in Allan Menzies (ed.), *Ante-Nicene Fathers*, vol 3: *Latin Christianity: Its Founder, Tertullian* (Grand Rapids: Christian Classics Ethereal Library, www.ccel.org/ccel/schaff/anf03.toc.html). Accessed July 23, 2010.

Chapter 15

In the Beginning and in the End

H. Peter Steeves

This is a confession.

We come to an end—as is the nature of things—and we look for meaning in it all, meaning that will somehow speak to the intricate relationships among religion, ethics, and the environment, carving out a place for us in the mix. We are confronted with the end, and we hope for a forever. My goal here is, perhaps, less ambitious if even more abstract—and this is the first part of my confession.

We could speak of covenants. Of the covenant God makes with humans before the flood and the possible reasons He expands this to a covenant with *all of creation* after the flood. We could speak of the Eucharist and what it means to celebrate communion with God, to celebrate transubstantiation, when we live in a world in which so many are hungry and our food—grown, engineered, and processed by distant corporations—is itself unhealthy. We could try to construct an ethic that transcends stewardship, ask whether or not there is any possibility for ecological justice without social and economic justice around the world, and try to find a way toward a Green Christianity. But I do not profess to have answers to the questions that all of this talking would raise. That is, in part, why I do the opposite. Confess and pro-fess may have begun as the same thing: acknowledging something *together* (con-) as well as *before* (pro-) the community. But the pros and cons of acknowledging what I do not know are too much to weigh. I am, by profession, a professor of philosophy. This, and more, I will confess.

I want to think about life. Its origins and its ends. The sense in which it stands at the center of Scripture but at the margins of the universe. And, perhaps most importantly, why we must move beyond our focus on the significance of life in establishing an ethic for *all* of Creation—an ethic that encompasses humans and animals; plants and microbes; rivers, rocks, and worlds; the animate and the inanimate. In keeping our eye on a universal morality, a story that will be true forever, we have privileged life. Forever? It is time to move on.

True community is constituted not only by family, friends, and the other humans with whom I share my life. It is made up of the variety of animal, plant, and nonliving natural things around me as well. And as such, we are all connected in a communitarian ethic, all inter-defining each other, creating and limiting freedom in every way, sharing our goods. It is something that St. Francis knew well, calling out in praise of the air and water, Brother sun and Sister moon, the elements of creation. Loving one's neighbor comes to include, for Francis, the animate and the inanimate—an ecological ethics comes to be the finest manifestation of a Christian ethic. But this is, still, and at least in metaphor, an ethic that anthropomorphizes or at least sees the whole of the world as living. The inanimate, that is, only becomes worthwhile in terms of seeing it animated. What I am interested in pursuing is the sense in which the inanimate, the unliving, is good in and of itself, has worth and meaning not as a mere backdrop and not as a precondition for life but as worthwhile in its very lifelessness.

What, we might ask, would it mean to be in community with and find inherent value in a rock? Just as it is possible for animals and plants to have goods, so too might we speak of the good of rocks—though to do so we will have to expand our understanding of goods and of ourselves.

We might, initially, look to see if phenomenology can give us a straightforward path to such an ethic. Is rock phenomenology possible? Animal phenomenology is possible, I have argued for some time now, because we are fundamentally animals.[1] There is no need to wonder "What is it like to be an animal" because I am always already animal. And so, can I practice rock phenomenology as well? Am I always already rock?

I make rocks. I have had three kidney stones in my life—though to say "had" already biases the investigation, as if the rocks were not a part of my body, a part of me, but rather something Other. Minerals are necessary to my very being; I am them. I am the iron in my blood, the magnesium in my brain. My body is 65 percent oxygen, it is true, and carbon, hydrogen, and nitrogen make up most of the rest. But the remaining 4.4 percent is what truly keeps me alive—the salts and trace elements that are essential for homeostasis. The 0.3 percent of my body that is sulfur is found in most of my proteins, and the 0.4 percent that is potassium is responsible for the continual beat of my heart. Limestone is "living" rock—and flesh, too, may become fossil. If I look carefully, all around me are the indications that I am enmeshed in a complex rocky community. When I think, I sometimes think like a rock. Aldo Leopold wrote that he tried hard to think like a mountain. Sometimes, I think he got such thinking almost right.

And so the first thing we might say is that of course rocks have goods, and this is evidenced by the fact that I have goods and I am rock. Some of my goods are particular to the parts of my life that are not rock-like, to be sure. Reading and writing don't seem to be things in which rocks are particularly interested. But the fact that I now drink two liters of water a day in order to ease the movement of any kidney stone that may form, the fact that I sometimes like to roll around and other times like to sit still, and the fact that I often prefer silence to speech—these are qualities that allow me to pursue rocky ends that I think I might have in common with other rock-beings. Still, though, there is much more to be said.

Epicurus agreed with Aristotle when it came to the human *telos* of pursuing *eudaimonia*. But the good, for Epicurus, was simply that which is most pleasurable.

---

[1]   See, for instance, H. Peter Steeves, *Animal Others: On Ethics, Ontology, and Animal Life* (Albany: SUNY Press, 1999) as well as Chapters 1–3 of *The Things Themselves: Phenomenology and the Return to the Everyday* (Albany: SUNY Press, 2006).

His argument is fairly phenomenological: pleasure is experienced straight-off as good; what other proof does one need?

Can a rock participate in *eudaimonia*? For Aristotle, of course, this is impossible, for rationality and *eudaimonia* are closely linked. What would it mean, though, for a rock to flourish, to have a good? For most people this is a nonsense question. How could a rock be worse or better off?

We might be tempted to say that what is good for a rock is that the rock be allowed to continue being what it is. Being ground away by blasting sand or melted in a passing lava flow would thus be bad. But this is to posit both a fixed, essential identity for the rock—as if there is something that it purely is and cannot be otherwise—and it is to reject change as ever possibly good. It is not enough, then, to say that we could identify the good of a rock in the rock being left alone.

Of course, for Aristotle rocks do have something *close* to "desires." They wish to be as low as possible. That's why they want to roll down hills rather than be pushed up them. As every Sisyphus-in-training knows, rocks may not have *soul* or an internal unifying force in a strict Aristotelian sense, but they do have an end-goal that is sometimes not in common with our own desires for them. Aquinas uses this as his example of a natural object's ultimate goal: rocks wish to be at the center of the universe (which is, of course, the center of the Earth). It is obvious simply by looking: rocks fall down not up, and when they reach the ground they continue to press down with force because they still want to end up someplace else, someplace lower still. A rock at the center of the Earth would thus simply float in place, a body with no force on it whatsoever. It would be, in some sense, a perfected rock—or a perfectly happy rock.

Now this is not meant to be an argument for why we need to mount an ethical project to take rocks back to the center of the Earth, though there has been, we can admit, a rock diaspora. Thanks to volcanoes, quarries, mountain upsurges and the like, rocks have been taken far from the depths of the Earth and scattered across the face of the planet. But this thinking is, instead, to get us considering ways in which a rock might indeed be better or worse off.

I am better or worse off based on how well I can pursue a virtuous life, how well I can reach my potential and flourish (which ultimately involves being a member of a healthy community). A chimpanzee can be better or worse off based on how well he can live his own life, avoid great suffering, become a functioning member of a community, and generally enjoy being alive. An acorn can be better or worse off based on how much water it gets, how the nutrients in the soil get fixed, and whether or not it can turn into an oak tree and become a functioning member of an ecosystem. And a rock can be better or worse off based on how well it … well … rocks. I'm not sure what this all entails. We have to think it through and get to know rocks in a much deeper way. But it certainly will include some measure of being a functioning member of a community, a landscape, an ecosystem, or something of the sort. It will involve some measure of integrity as well, but not one that makes us demonize all change. It might even have something to do with being respected by others.

This is more than just a rejection of utilitarianism and Kantianism, then. It is a hope that we can found some deeper sense of the moral by means of thinking through what it truly means to be together, something that transcends rights and *hedon* point calculations. Something that overcomes the language of individuals, subjects, and moral agents and patients—that overcomes a language that stacks the deck against certain things mattering from the very start.

Starting in the West, perhaps, with Plato, we need to rethink our world. In the allegory of the cave, the prisoners are said to live a life where all they know are shadows. Come to know the things themselves, Plato demands, not the shadows of the things on the cave wall. What this overlooks, though, is that the prisoners have indeed already known something: the cave-wall itself, the rocky slab before them. The rocks, the world, are not a backdrop against which the passing parade of more interesting things flicker by. Let us, then, reclaim Plato's myth in the name of the rocky cave wall! Being-in-the-world means being in a place. And this means that we must know the world itself: the humans, animals, plants, rocks, and shadows. Before the freed prisoners leave the cave and get blinded by the sun, I put it to them that they should get to know the cave walls even better.

No doubt I have not defused a skeptic's worries about an ethic for nonliving things by focusing briefly on rocks here. I have probably only dug my hole deeper (this is another trait I share with rocks: we like being at the bottom of holes). And if I continue to think about why life itself is unimportant in the grand scheme of things, I will perhaps only make matters worse, veering away from philosophy and from religion for a moment. But we have to do it anyway.

The simple fact is that I hold out hope that there is meaning in the universe that transcends the human tendency to attach meaning to meaningless things. I believe that there is a purpose, a reason, a *something* behind the cosmos, a something behind the *something* rather than the *nothing* that exists. An answer to the biggest *why* question imaginable. But if that is the case, then we have to accept the fact that life is likely not a major part of that answer. I love life. And I love living things. But I worry that as ethicists we have fallen into a bias: we are lifeists. And like sexism, racism, classism, and speciesism, *lifeism* must be overcome.

I do not escape this obsession with life myself. I recall crying in Sunday school at the age of six when I finally came to understand just what it meant to say that Jesus had *died*. I have faced my own near-death several times and, even worse, the actual-death of far too many others I have loved. And I will further confess just how far this obsession with life has taken me. In my secret life, my life beyond the academy, I am sometimes lucky enough to be a part of a community of researchers at NASA Ames—a community investigating the origin of life on Earth.

Life, it is sometimes said, seems to violate the second law of thermodynamics, the law that demands that entropy, a measure of disorder, is always on the rise. Life takes disorder and makes something—a living thing that is extremely orderly— from that chaos. This is a reversal of chaos, a decrease in chaos. It is, in the most Biblical sense, *genesis*. Out of the madness of the bubbling primordial soup, a little strand of life supposedly appears: order from nothing, a violation of the most basic law of nature. It seems, then, almost like a miracle.

We know, of course, that little pockets of order can arise in nature when this is the easiest means toward fulfilling the second law and increasing entropy in the world. Perhaps life, then, is merely the most efficient means of destroying some orderly gradient in the universe? Perhaps life must arise at a certain moment when conditions are just right in a certain age of the cosmos?

Perhaps. Yet in the lab we have been unable to reproduce that moment. Unable, even, to get close. The search for the origin of life is no further along than the search for the Holy Grail. And so, where do we go from here? One promising direction, I believe, involves reconceptualizing our experiments and our definition of "life" itself. That is, what if "life" is being defined in an overly mechanistic, secular way that strips it of its teleological essence and thus life is never showing up in the lab because we simply have misunderstood what it means to be alive?

To paraphrase Aristotle, "life" is said in many ways. Most attempts to define "life" run aground on familiar shores. Common to most of them, however, is a sense of self-made destiny. Against Aristotle, there is little teleology these days in our understanding of life, unless that teleology is itself self-constituted. We must call this into question and think about what is lost when teleology is lost. Perhaps the universe, unlike God, did not mean to create life—it certainly did not *mean* to create humans—but in tossing out all talk of meaning in our discourse, we overlook the possibility that the definition of "life" might best be had by looking at what it *does* rather than what it *is*, looking at what purpose it can serve rather than what parts make it up. Even the most hearty and strapping definitions of "life" are ultimately too boot-strappy and too without heart. And this is a consequence of historical accident, not scientific necessity.

To take some traditional definitions, we might say that a thing is alive if it is autonomous, if it metabolizes, if it reproduces itself. Immanuel Kant saw the difference between a machine and an organism to be that in a machine the parts exist *for* each other but not *by* each other; their operation has nothing to do with building the machine. The parts of an organism work together but also produce the organism: each part is cause and effect, and in this sense an organism is, for Kant (and in modern parlance), a self-organizing entity. More contemporary definitions give their own twist to this basic template. Lynn Margulis and Dorion Sagan see life as that which is autopoietic. From the Greek for "self" (*auto*) and "making" (*poiein*), autopoietic life continually produces itself, engages in self-maintenance. Stuart Kauffman defines life as a self-reproducing system that is capable of performing at least one thermodynamic work cycle—an emergent property in an autocatalytic chemical reaction network. And biochemist Johnjoe McFadden takes life to be that which is made of a set of peptides using internal quantum self-measurement to capture low-entropy states in order to put off thermodynamic decay.

It's not that I wish to argue against any of these definitions, but rather to point out some of the underlying assumptions that may have us looking, in the end, in only one direction. Life is taken to be something special in most definitions, something that is made from nonliving parts yet as a whole is somehow self-made. The words "auto," "self-making," "self-measuring," and "self-reproducing" are the key terms. And they are all reflexive. Life becomes life by doing something to itself.

Philosophy has made a career of studying self-referentiality. For 2,500 years, Western thought has puzzled over—and turned in a crisis to—self-reference. There seems to be something almost magical about it. It defies logic and appears on the scene to get us out of all sorts of binds. It is both trickster and savior.

In ancient Greece, the liar's paradox was well known. Whenever a liar claims something to be true, it is easy to know that it is in fact false. Unless the liar claims "I am lying." When the liar refers to himself in a sentence that refers to itself, all hell breaks loose.

Seeking to tame the self-reference, René Descartes turned it into the foundation for all epistemology in his search for apodictic knowledge. The inability to doubt the fact that he was doubting would put philosophy on a solipsistic epistemological path we are still struggling to overcome. Edmund Husserl updated Descartes nearly three centuries later, arguing that the mind is always directed, and that if we wish to unlock its very structure, we need only bend consciousness back upon itself thus making the object of consciousness consciousness itself. By undertaking a phenomenological epoché, we can consequently come to investigate intentionality. And Jacques Derrida, too, would later go on to claim that everything is ultimately self-referential because there is nothing outside of the text. We are all caught up in a hermeneutic of being and can find no place *outside* to comment objectively on anything.

Philosophers outside of the continent also are preoccupied with self-referentiality, though in a more analytic way. Studying the logic and the language of self-reference, Gödel discovered that given any axiomatic system as or more thorough than arithmetic, there will be some statements that are true while at the same time cannot be proven true (cannot be derived from the axioms). These special statements are, of course, self-reflexive—as interesting set-theory statements often are. One way, crude but in the right spirit, of understanding Gödel's insight is trying to imagine a book that would compile all books within its covers. It would have everything Gabriel García Márquez ever wrote, every edition of Shakespeare, all of James Joyce, every book ever published anywhere. Yet it would always be incomplete because once all other books were finally compiled, it would still be missing one book—namely, *itself.* And if we were to put a copy of itself inside its covers (thus doubling the contents) it would still be incomplete, because now there would exist a new book (the newly-expanded self edition) that would still be missing.

The move to self-referentiality is always done with flourish. It marks the demise or the salvation of a system of thought, but always something importantly system-shattering. Our preoccupation with self-reference, though, has its historical roots.

And it rises from a preoccupation with the self. One might even call it an idolatry of the self.

"Autonomy" is neither an ancient nor a natural term. Seeing things as separate, self-sufficient, and self-directed has a political and metaphysical history. In the seventeenth century this radical individualism took a secure hold on our culture in the form of Liberalism. Descartes' isolated monadic self would be the start of all modern Western metaphysics and epistemology. Thomas Hobbes' isolated, selfish, and equal creatures warring in a state of nature would found all social contract theory, which is to say most political theory in the West. Galileo and Newton's mathematization of the natural world and conception of objectivity as the opposite of subjectivity would set the standard for scientific inquiry. The Other subsequently becomes a nuisance in the search for truth; the Other is at best an afterthought. Philosophy consequently undertakes a centuries-long project of proving the existence of other minds; democracies spring up in which I am most free when I am most left alone. Communities—being inherently with Others—are vilified. And thus when we go to define "life" we "naturally" start by seeing whatever is alive as radically individual, a unit, an autonomous thing; and we see the spark that brought it into being as necessarily self-given. For what else is there apart from the self? Life, for the Liberal, is an individual enterprise.

I recall watching Carl Sagan on TV in my youth pour the elemental ingredients of a human body into a large glass box—mostly water, of course. He stirred it up, smiled in that way that he smiled on his PBS show, and asked why nothing was alive in there. The same sorts of questions had already been haunting me for years: what are we, where did we come from, what makes us alive, how did the cosmos begin, and what is this all for? My childhood answer to Carl's question about animating the stuff of life went back and forth between "soul" and "electrical-chemical energy." These seemed to me the two possible poles, divine and secular, and I flitted back and forth from one to the other with youthful life-force and indecisiveness.

The image has stayed with me all of these years: what was missing in that box? Order, of course. One cannot put the parts of an airplane on a runway and wonder why nothing takes off. The Being of the pieces is also defined in terms of their being-together. But it seems to me today that there might be something more deeply problematic with the question itself. The chemicals that Carl stirred in his box were of just the right proportion to make a single human. And one way or another I fear we are still searching for how that one, single, solitary, unitary, autonomous individual came to life: the first cell, the first strand of DNA, the first self-replicating peptide. And this is all very Liberal.

I wonder: could the first living thing have been *one* thing and still have been alive; what would it mean to be *one* living thing? Robinson Crusoe makes sense as a story because we in community read it, we in community imagine someone stranded from community, someone thus dangerously close to losing his humanity, his identity. We tell stories of feral children and even have some factual cases; and in each instance the isolation has rendered the being unrecognizable.

What would it mean to be the *first* living thing? Without a biosphere, without an ecological environment, without an Other? When we imagine the first living thing, is our Liberalism shaping our experiments and our models? Are we looking for a Hobbesian peptide in a state of nature, a Cartesian RNA-self in radical isolation in a little pond, a Leibnizian mound of monads in the bubbles atop a four billion year old percolating ocean? Is the myth of the capitalist's self-made man rewritten as the myth of the first self-made molecule, pulling itself up by its own bootstraps to make it in the big-time game of Life like a mitochondrial Bill Gates?

The Biblical account of genesis is one in which nothing is ever alone. On the first day God makes the heavens and the Earth together. He goes on to make the plants and the beasts in communities, never one by one. Jesus' ethic is communitarian as well. We love our neighbors, care for the marginalized, strive for practical justice. The Christian ethic, like Aristotle's, is phenomenological at heart. Not so much a listing of rules and orders and laws, but a description of how good people act. Not so much a striving for some Heaven that is after life, but an acting as if the Kingdom of God is already spread upon the Earth. It is such action, of course, that makes the Kingdom appear, but this is only a paradox for someone who has never read German philosophy. Jesus calls on us to live as if we are in paradise, as if we and the Earth are already saved.

We are in this together. Human sin and environmental death are found everywhere together in the Bible. The Fall brings about the end of a peaceful relationship between humanity and the rest of nature. Cain's curse, after slaying his brother, is that the Earth will no longer support him. God destroys the sin of the world with the baptismal Flood that destroys the world itself. Everywhere—from Gen 7:21–22, Lev 18:25, Deut 29:22, Amos 4:7, and all throughout Revelation—the fate of the Earth and the fate of humanity are tied together. But this does not mean that we must think of ourselves as stewards or as outside of nature. We are not the final product of creation.

Indeed, the first creation story in Genesis does not have humans as the pinnacle of creation but rather God's resting. It is all leading to the Sabbath, that is, to an *absence* of work. This is where the story is headed. Not toward the creation of the heavens or the fish or the humans, but toward resting. This experience of the absence of labor shows itself in Genesis as a present-absence. It is not a nothingness, but a quiescence, a fallow-time, an active refrain from activity. And so the priests know: the void does not precede the universe, the resting does not precede the labor, death does not precede life. The experience of absence is as a presencing. And the point toward which all creation is moving is toward the Sabbath. This is the *telos*, the final state, the goal. The universe coming to rest. And we who are alive dream of forever. A forever that somehow includes us. A forever that somehow favors us.

Here, in conclusion, is our conclusion, our final resting place. A final scientific story: a brief history of the past and the future of the universe. Let us confess this together.

In roughly six-and-a-half billion years our sun will be going through its final death throes and will have swelled into a Red Giant, totally engulfing Mercury and Venus. At the Earth's equator the temperature will reach 3,000 degrees Fahrenheit, and the sun's enormous sphere will fill up more than half of the sky at noontime. A few centuries after this, the Earth itself will disappear as the sun expands to our orbit and absorbs the planet, continuing part of the way out to Mars before cooling and collapsing back on itself to die. Life, of course, will not have been possible on Earth for millions of years before this. There will be no one, save the molten bodies of stones, to witness the end. In the lifespan of the universe, all of this takes place in the blink of an eye. A trillion trillion trillion more years will still come to pass. And then a trillion trillion trillion more. And we search for meaning in it all, hoping for forever.

But the fact that I will die and you will die and our planet will die and our sun will die means, of course, nothing to the universe as a whole. Life itself—not just human life or plant life or alien life, but life in general—is just a passing fad in the universe. It is a blip in the cosmic timeline, only worth mentioning become we are in the life phase of the universe—short as it is—right now. And it is, indeed, very short.

Moments after the Big Bang the temperature of the universe is far too hot for any matter to exist at all. In the first 10^{-35} seconds, space swells from a size smaller than a dot to a size larger than the visible universe today. This inflation quickly slows down, though it continues on in one form today, and the temperature drops enough to form the building blocks of matter.

In this next phase of the universe, at the young age of one microsecond, quarks and anti-quarks arise and are tossed about in a seething sea of radiation, annihilating each other. After only 30 microseconds there are no free quarks left at all in the universe. The minute excess of matter over anti-matter in the early phase of the universe gave us a matter-based cosmos, and when the universe finally turns one second old, protons and neutrons begin combining to form helium and hydrogen nuclei. For a few minutes the universe makes helium—nearly all of the helium that exists today was formed during those first three or four minutes of the universe. It's a helium party. And the party is a gas. And afterwards, for roughly 300,000 years, nothing much changes apart from the gradual cooling and expansion of space.

When the cosmos finally cools to 3,000 degrees Kelvin, stable atoms of hydrogen are able to be formed from the free-floating hydrogen nuclei and electrons. Ordinary atomic matter, of the type we are used to, starts to take root. And as that matter has mass, is cooling, and is slowing in its movement, gravity begins to have its effects felt in the cosmos for the first time. Hydrogen attracts hydrogen, and atoms begin congregating. Galaxies composed of clouds of hydrogen and helium begin to form, and within them the first stars are born. Thus begins the Stelliferous Era of the universe. Suddenly there are spots of light in the sky. Planets arise from the accretion disks around stars several hundred thousand years later. Solar systems are born. The stars light up and shine their light down on planets. And thanks to the heavy elements created by early stars undergoing supernovae, some of these planets are filled with exotic things such as carbon and nitrogen and oxygen: the building blocks of life.

Life arises on some of these planets and perhaps in interplanetary space when possible. A little bit less than 14 billion more years pass, and you end up on the Earth reading these words. It's grand. We have trees and lilacs and lightning bugs and cornfields. We have a paradise of life. And we think to ourselves that this is what it's all about: the universe set out to make us, or if it didn't set out to make *us* exactly at least it set out to make life. Life is what it all means. We must restore it when it's gone, worship it while it's here, and pity the beings that do not have it and thus fall outside our ethic.

But life is just another passing fad, a phase soon to be over.

Just as our star will die and engulf the Earth, so every star in the cosmos will eventually die. Our sun will die soon in the grand scheme of things. The Stelliferous Era has a long way to go after our solar system perishes. But about one hundred trillion years from now, the final stars in the universe—every last one of them—will die out. No more supernovae, no more exotic elements, no more life. This is nearly the end of the things themselves. But the universe's story is still far from over.

In the years that follow, protons start to decay. The lifespan of a proton is ten trillion trillion trillion years, but as the protons die, so dies the possibility of congruent atoms. We leave the phase of stars and the phase of life, and enter the phase during which black holes reign. Sucking up more and more bits of ripped-apart matter, the black holes of the universe grow more and more massive. After 10^{40} years, there are no more galaxies in the universe, only the large black holes that sit at the center of what used to be their host galaxies. As the black holes approach each other—moving like impossibly slow, insatiable beasts—they sometimes collide to create super-black holes. Little by little these leviathans vacuum up all of space leaving nothing but themselves behind. Space goes completely black.

Trillions upon trillions of years pass. But the end has still not yet arrived, for as massive and as ravenous as black holes are, they have a singular problem: they leak. Hawking radiation steadily flows out of a black hole, the result of quantum effects, which means that little by little, resting black holes lose their mass. The process is incredibly slow, but it is also incredibly inevitable. Even black holes must come to an end. As a leaky black hole finally reaches a tiny fraction of its original mass, it explodes and lights up the universe one last time. At 10^{67} years old, the cosmos starts to fill with these little crackles, like paparazzi flashbulbs popping and quickly extinguishing across the universe. When they are done, there will never be any more light, ever.

At some point after 10^{100} years, an unimaginably long time, it is over. Everything is cool, nothing is moving, what little bits of matter that remain are too far apart for gravity to have any effect. The maximum entropy state of the universe is reached. Chaos wins the battle against order. The ultimate thermodynamic heat-death of everything comes to pass. The universe stops; time comes to an asymptotically infinite end. And life—long, long gone—is barely even a footnote worth mentioning.

Here. Here is your forever.

Bibliography

Steeves, H. Peter, *Animal Others: On Ethics, Ontology, and Animal Life* (Albany: SUNY Press, 1999).

Steeves, H. Peter, *The Things Themselves: Phenomenology and the Return to the Everyday* (Albany: SUNY Press, 2006).

Index